GREENING EXISTING BUILDINGS

McGRAW-HILL'S GREENSOURCE SERIES

Gevorkian
Solar Power in Building Design: The Engineer's Complete Design Resource
Alternative Energy Systems in Building Design

GreenSource: The Magazine of Sustainable Design
Emerald Architecture: Case Studies in Green Building

Haselbach
The Engineering Guide to LEED—New Construction: Sustainable Construction for Engineers

Luckett
Green Roof Construction and Maintenance

Melaver and Mueller (eds.)
The Green Building Bottom Line: The Real Cost of Sustainable Building

Nichols and Laros
Inside the Civano Project: A Case Study of Large-Scale Sustainable Neighborhood Development

Yudelson
Green Building Through Integrated Design
Greening Existing Buildings

About *GreenSource*
A mainstay in the green building market since 2006, *GreenSource* magazine and GreenSourceMag.com are produced by the editors of McGraw-Hill Construction, in partnership with editors at BuildingGreen, Inc., with support from the United States Green Building Council. *GreenSource* has received numerous awards, including American Business Media's 2008 Neal Award for Best Website and 2007 Neal Award for Best Start-up Publication, and FOLIO magazine's 2007 Ozzie Awards for "Best Design, New Magazine" and "Best Overall Design." Recognized for responding to the needs and demands of the profession, *GreenSource* is a leader in covering noteworthy trends in sustainable design and best practice case studies. Its award-winning content will continue to benefit key specifiers and buyers in the green design and construction industry through the books in the *GreenSource* Series.

About McGraw-Hill Construction
McGraw-Hill Construction, part of The McGraw-Hill Companies (NYSE: MHP), connects people, projects, and products across the design and construction industry. Backed by the power of Dodge, Sweets, *Engineering News-Record* (*ENR*), *Architectural Record*, *GreenSource*, *Constructor*, and regional publications, the company provides information, intelligence, tools, applications, and resources to help customers grow their businesses. McGraw-Hill Construction serves more than 1,000,000 customers within the $4.6 trillion global construction community. For more information, visit www.construction.com.

About the International Code Council
The International Code Council (ICC) is a nonprofit membership association dedicated to protecting the health, safety, and welfare of people by creating better buildings and safer communities. The mission of ICC is to provide the highest quality codes, standards, products and services for all concerned with the safety and performance of the built environment. ICC is the publisher of the family of the International Codes® (I-Codes®), a single set of comprehensive and coordinated national model codes. This unified approach to building codes enhances safety, efficiency, and affordability in the construction of buildings. The Code Council is also dedicated to innovation, sustainability, and energy efficiency. Code Council subsidiary, ICC Evaluation Service, issues Evaluation Reports for innovative products and Reports of Sustainable Attributes Verification and Evaluation (SAVE).

Headquarters: 500 New Jersey Avenue, NW, 6th Floor, Washington, DC 20001-2070; District Offices: Birmingham, AL; Chicago, IL; Los Angeles, CA, 1-888-422-7233, www.iccsafe.org.

GREENING EXISTING BUILDINGS

JERRY YUDELSON, PE, MS, MBA, LEED AP

New York Chicago San Francisco Lisbon London Madrid
Mexico City Milan New Delhi San Juan Seoul
Singapore Sydney Toronto

The McGraw-Hill Companies

Cataloging-in-Publication Data is on file with the Library of Congress.

McGraw-Hill books are available at special quantity discounts to use as premiums and sales promotions, or for use in corporate training programs. To contact a representative please e-mail us at bulksales@mcgraw-hill.com.

Greening Existing Buildings

Copyright ©2010 by The McGraw-Hill Companies, Inc. All rights reserved. Printed in the United States of America. Except as permitted under the United States Copyright Act of 1976, no part of this publication may be reproduced or distributed in any form or by any means, or stored in a data base or retrieval system, without the prior written permission of the publisher.

1 2 3 4 5 6 7 8 9 0 DOC/DOC 0 1 4 3 2 1 0 9

ISBN 978- 0-07-163832-6
MHID 0-07-163832-6

The pages within this book were printed on acid-free paper containing 100% postconsumer fiber.

Sponsoring Editor
Joy Bramble Oehlkers

Acquisitions Coordinator
Michael Mulcahy

Editorial Supervisor
David E. Fogarty

Project Manager
Somya Rustagi, Glyph International

Copy Editor
Megha Roy Chowdhury

Proofreader
Shivani Arora, Glyph International

Indexer
Jerry Yudelson

Production Supervisor
Richard C. Ruzycka

Composition
Glyph International

Art Director, Cover
Jeff Weeks

Information contained in this work has been obtained by The McGraw-Hill Companies, Inc. ("McGraw-Hill") from sources believed to be reliable. However, neither McGraw-Hill nor its authors guarantee the accuracy or completeness of any information published herein, and neither McGraw-Hill nor its authors shall be responsible for any errors, omissions, or damages arising out of use of this information. This work is published with the understanding that McGraw-Hill and its authors are supplying information but are not attempting to render engineering or other professional services. If such services are required, the assistance of an appropriate professional should be sought.

About the Author

Jerry Yudelson, PE, MS, MBA, LEED AP, is the Principal of Yudelson Associates, a green building and sustainable planning consultancy based in Tucson, Arizona. He holds engineering degrees from the California Institute of Technology and Harvard University, as well as an MBA (with highest honors) from the University of Oregon, and he is a licensed professional engineer (Oregon). Mr. Yudelson has spent his professional career engaged with energy and environmental issues, and has been involved on a daily basis with the design, construction, and operation of residential and commercial green buildings. He works for architects, engineers, developers, builders, and manufacturers to develop sustainable design solutions. His work on design projects involves early-stage consultation, eco-charrette facilitation, and providing LEED expertise and coaching for design teams. He works with developers and building teams to create effective programs for large-scale green projects, as well as with product manufacturers to guide them toward sustainable product marketing and investment opportunities. In addition to this general business and professional background, Mr. Yudelson served for 8 years as a LEED national faculty member for the U.S. Green Building Council (USGBC). Since 2001, he has trained more than 3500 building industry professionals in the LEED rating system. He has served on the USGBC's national board of directors and, since 2004, he has chaired the steering committee for the USGBC's annual conference, *GreenBuild*—the largest green building conference in the country. He is the author of *Green Building through Integrated Design; Green Building Trends: Europe; Sustainable Retail Development: New Success Strategies; Green Building A to Z: Understanding the Language of Green Building; The Green Building Revolution; Choosing Green: The Homebuyer's Guide to Good Green Homes;* and *Marketing Green Building Services: Strategies for Success.*

CONTENTS

FOREWORD

I came across a Web site the other day with a number of fun T-shirt slogans:

- I'm a PB&J sandwich in a caviar world.
- Did you ever think you were a Raggedy Ann doll in a Barbie doll world?
- I'm a glass of tap water in a grande-soy-no-foam-vanilla-latte-served exactly at 153.2-degrees world!

And my favorite: Did you ever feel like the world was a tuxedo, and you were a pair of brown shoes?

I know that last one came from the late comedian George Gobel. I sometimes think about it when talking about LEED for Existing Buildings: Operations & Maintenance (LEED-EBOM). People get all excited about the flashy new buildings constructed and certified to LEED for New Construction (LEED-NC) standards. And believe me, they're beautiful!

But if LEED-NC is the tuxedo, then LEED-EBOM is the pair of brown shoes: enduring, functional, and comfortable, not flashy. And I'm okay with that. LEED-EBOM provides the deep dive into our nation's plethora of existing buildings. It's literally the nuts and bolts of sound operations, including plumbing fixtures, solid waste management, and most important—building commissioning. By providing documented confirmation that building systems function according to the initial criteria, not only are savings achieved, but building owners and occupants are happier. Using commissioning and all the other LEED tools, buildings as diverse as the Owens Corning campus in Toledo and the Getty Museum in Los Angeles are showing the world the power of ingenuity.

It has been great to find a kindred spirit in Jerry Yudelson. He will tell you about many best practices of LEED-EBOM in this book and help you become more familiar with why this information is so urgently needed, as our world faces unprecedented economic, environmental, and social challenges. The opportunities are exciting, as colleges, hospitals, governments, corporations, and others finally have the tools they need to take a comprehensive approach to their entire building portfolio. You'll find out how they're balancing the business case for greening existing buildings with the costs, and how dealing with such boring old brown-shoe topics as motor controls and mechanical systems is generating exciting new technologies.

One of the most valuable lessons, especially in tough economic times, is the discussion of how funding methods such as performance contracting can make it all possible. Energy service companies (ESCOs), including Johnson Controls, have used performance contracting frequently in the public sector. It's a way to invest in new equipment with the costs repaid by the accrued savings in energy, water, and operations, and the ESCO guarantees the work. Our performance contracts involve large public projects, such as the Oak Ridge National Laboratory where we recently signed a $89 million contract to reduce energy use by 850 million Btus (250,000 kWh) per year, reduce water consumption by 170 million gallons (643 million liters) a year, and develop a large biomass gasification system for the campus, as well as smaller jobs such as municipal facilities in Laurel, Montana.

Now the private sector is putting its best foot forward, beginning with one of the biggest existing buildings of all, the Empire State Building. The initial $20 million renovation project announced in April 2009 will result in the Empire State Building annually reducing energy use by 38 percent and energy costs by $4.4 million. Internal calculations show that the Empire State Building will be able to qualify for LEED-EBOM Gold certification.

The project involves a coalition of leading organizations focussing on sustainability, including Johnson Controls, Jones Lang LaSalle, and the Rocky Mountain Institute, all brought together with the support of the Clinton Climate Initiative. Johnson Controls is guaranteeing the energy savings via a performance contract, which expects to obtain third-party energy-efficiency financing for retrofit projects. Best of all, the entire process is available online as open source materials for public use, at www.esbsustainability.com, so others can learn from the experience.

An important element of the Empire State Building project and others is teaching tenants about the facility improvements and the impact their behavior can have on reducing costs. In my experience, successful projects always involve a wide variety of people. It starts with the top—perhaps a company chairman or university president—and relies on facility managers and maintenance engineers. It also involves collaboration and communication with tenants, students, sales, marketing, and every other person in an organization or company—as well as the outside community.

Because if there's anything I've learned in several decades of working in energy and the built environment, it's that we're not just fixing buildings. We're transforming organizations, converting communities and, dare I say it, changing the world.

Sometimes the impetus is from the boiler room; sometimes it's from the boardroom. The first step, beginning with books like this and of course the Internet, is education: learning what works and what does not.

A combination of these efforts has led to sustainable success for Johnson Controls itself. Our approach involves people working individually and collectively to transform their coworkers, their organizations, and the communities they serve:

- Our CEOs have been absolute in their commitment to sustainability and quality, its close associate. We require our three business units to report on 10 key sustainability strategies every year, including working with our suppliers to improve the eco-efficiency of the supply chain and ensuring the creation and expansion of green jobs.
- While our success is directly related to executive leadership, it's also a factor of strategic support. Ward Komorowski, our director of facilities, was behind the huge job of helping Johnson Controls Brengel Technology Center become one of the first

Figure F.1 Owned by Johnson Controls, the Brengel Technology Center in Milwaukee, Wisconsin, was the first LEED-EB Gold project awarded for a single-tenant corporate building. *Photo courtesy of Johnson Controls, Inc.*

buildings in the country to receive LEED-NC in 2001 and the very first to be recertified under LEED-EB in 2004. Now, Ward is tackling his most ambitious undertaking: helping direct construction of a new Johnson Controls corporate headquarters campus, designed to receive LEED-NC Platinum.

- As the Johnson Controls organization continues to transform, we're practicing our third element of success: collaboration and communication. Through our Blue Sky Involve program, we share expertise across business units, offer leadership training, and use a wide variety of internal and external channels.

We've learned that to really engage people and develop a culture of sustainability, it's important to educate and communicate throughout the process, not just when you put a plaque on a building. And that's how LEED-EBOM will have the biggest influence. As people in organizations such as the Natural Resources Defense Council and ICLEI—Local Governments for Sustainability work to expand the impact of green existing buildings to neighborhoods, cities, counties, and states, they truly can make a difference when they follow many of the best practices described in this book.

I've seen the profound impact many people have had in advancing sustainable design. In fact, I've seen many of them in formal attire as they accept awards on behalf of their work. But, like me and many readers of this book, I suspect most of them are even happier wearing respectful, reliable, resourceful brown shoes as they simply go about changing the world.

Paul von Paumgartten
Director, Building Efficiency Energy and Environmental Affairs
Johnson Controls
Milwaukee, Wisconsin

PREFACE

I've been writing books about green buildings, green homes, and green developments since 2005; each has focused mostly on new buildings, with the objective of helping building owners, architects, developers, contractors, engineers, and homebuilders to understand and implement the business case for green building.

Green building growth appears now to be self-perpetuating, expected to achieve nearly a 20 percent market share of the new nonresidential construction market in 2009. Now, the time has certainly come to focus on existing buildings, which after all represent the great majority of all buildings and which contribute approximately 20 percent of all U.S. greenhouse gas emissions.

Similar conclusions hold true for other developed economies: in Canada, western Europe, Japan, and other countries; existing buildings are where we must look to reduce greenhouse gas emissions and to ensure more livable places for people to work, live, play, and learn.

A seminal event, the election of President Barack Obama and an overwhelming Democratic Congress in late 2008, makes it almost certain that greening existing buildings will move up much higher on the scale of national priorities. Over the next four years, the United States, the world's largest economy and second largest generator of greenhouse gas emissions (after China), will begin to tackle seriously the challenge of human-caused global warming and attendant issues of ameliorating rapid climate change. There is no way to do this without addressing the energy use of existing buildings.

In many ways, the challenges of greening existing buildings are far greater than greening new buildings; we are not starting with a blank slate, as with new buildings, but with an existing edifice and set of operating practices. In many situations, it's not easy or cheap to change the building envelope, it may not be economical to change out the HVAC equipment, and a significant percentage of the building's energy use is already determined by scale, mass, and orientation.

Nevertheless, as this book demonstrates, there is still a lot we can do. On very large buildings, as you'll see with New York's iconic Empire State Building, it may even pay to replace all the windows. On many properties, the greatest savings will come from dozens of measures, individually humdrum, but collectively significant.

Beyond savings in energy, water, and waste management expenses, the real gains in greening existing buildings lie in the seemingly "soft" benefits: improvements in health, comfort, and productivity of building occupants; enhanced marketing and public relations; risk mitigation, improved recruitment and retention, and greater employee morale. As building owners look to make a business case for greening an existing building, whether a single-tenant or multitenant building, these other benefits constitute a strong part of the justification.

How should a building owner, building manager, or facility manager go about greening existing buildings? This book focuses mostly on using established green building rating systems, especially the U.S. Green Building Council's LEED for Existing Buildings (LEED-EB), but includes consideration of newer evaluation schemes for existing buildings from the U.K.'s Building Research Establishment's BREEAM rating system and the Green Building Council of Australia's Green Star program.

In this book, I have chosen to emphasize use of the USGBC's LEED-EB rating system, for three reasons. First, it is the longest established and most widely used of the existing building rating systems, having first appeared in a pilot program in 2002. Second, it offers a full range of options for greening existing buildings, measures duplicated in other rating systems, so that by examining projects using the LEED system, you will understand how to use the other rating systems in countries where they are more widely used. Third, LEED-EB is the fastest growing of all LEED rating systems since 2007, indicating a growing market acceptance of the system.

Having decided to focus on LEED-EB, I then decided to focus primarily on completed projects that have received the highest ratings in that system, the Gold or Platinum designations. Interestingly, most of the completed projects are by private owners, corporate-owned and operated properties, commercial buildings owned by smaller local enterprises, and commercial buildings managed by large national and international firms. That private owners are willing to incur the costs of upgrading their properties to the highest standards, alone speaks volumes about the benefits of greening existing buildings.

I have not neglected the universities with strong commitments to greening their campuses or the government agencies, federal, state and local, with similar commitments and achievements. After all, collectively, government activity represents more than one-third of the total U.S. economy; moreover, government agencies, colleges, and universities expect to own and operate their properties for decades (if not centuries) to come.

The most significant "father" of the LEED-EB program and its longtime champion, is Paul von Paumgartten, Director of Energy and Environmental Affairs for Johnson Controls, Inc. I recall a dinner meeting with Paul (or PvP, as he is widely known in green building circles) in Los Angeles in 1999, when he advocated passionately that the LEED rating system for new construction, then still early in its first pilot evaluation stage, should be expanded to include existing buildings. I asked this tireless (and very effective) advocate for upgrading the existing building stock to write the foreword for this book.

Greening Existing Buildings shows the way for anyone involved with building ownership and operations to upgrade the energy and environmental performance of almost any building. In the book, I focus on lessons learned in actual projects, profile more than 25 LEED for Existing Buildings certified projects and use interviews with more than 35 industry experts and building management practitioners. It's my fervent hope that you will take this information and put it to use in your own buildings, facilities, factories, hotels, hospitals, high-rise residences, schools, and campuses.

As a country, and as a world, we need to get moving quickly to dramatically reduce carbon emissions from the existing building stock. By profiling the large number of successful projects, by demonstrating the business case, and by showcasing the wide range of specific strategies for greening existing buildings, I hope to motivate you, the reader, to begin taking action now in your own spheres of influence.

JERRY YUDELSON
Tucson, Arizona
Sonoran Desert Bioregion

A Note on Nomenclature: In this book, I generally use the term LEED-EB for the versions 1.0 and 2.0 of the LEED for Existing Buildings rating system. Beginning in 2008, the rating system was renamed LEED for Existing Buildings: Operations & Maintenance, which I shorten in the book to LEED-EBOM. Sometimes, I just use LEED to represent all LEED for Existing Buildings rating systems. Where other LEED rating systems are discussed, such as LEED for New Construction, I use their full names and/or acronyms.

ACKNOWLEDGMENTS

First of all, I want to thank the more than two dozen green building, finance, and real estate industry professionals who allowed us to interview them for this book, supplied project information and have led the way in greening existing buildings. Those we interviewed are acknowledged individually in Appendix III, "Interviewees." I also want to thank my editor at McGraw-Hill, Joy Bramble Oehlkers, for championing this book and seeing it through to timely publication, and Kraig Stevenson, International Code Council Senior Regional Manager, Bellevue, Washington.

Thanks as well to the many companies, building owners, developers, architects, architectural photographers, and others who generously contributed project information and photos for the book. Thanks also to Heidi Ziegler-Voll for providing illustrations created specially for this book.

A special note of thanks goes to my in-house editor, Gretel Hakanson, for conducting the interviews, sourcing all the photos and permissions (a never-ending task), reviewing the manuscript drafts, and making sure that the final production version was accurate as possible. This is our eighth green-building book produced together; she has been an invaluable contributor to each work. A special thanks also goes to Yudelson Associates' research director Jaimie Galayda for preparing an earlier draft of Chapter 7, preparing much of the appendix material, gathering information used in various chapters, and reviewing the entire text; and to Todd Leber for drafting the case study write-ups in Chapter 13 and preparing the material in Appendix I.

Thanks to those who reviewed the book proposal and offered helpful suggestions, and to Professor Ulf Meyer, Erik Ring, and Joyce Kelly for their detailed reviews of the draft manuscript. As always, any errors of omission or commission are mine alone. Thanks also to my wife, Jessica, for indulging me as I spent time on yet another green-building book and for sharing my enthusiasm for sustainable living.

GREENING EXISTING BUILDINGS

1

THE SUSTAINABILITY REVOLUTION

All of humanity now has the option to "make it" successfully and sustainably, by virtue of our having minds, discovering principles and being able to employ these principles to do more with less.

—R. Buckminster Fuller, *Utopia or Oblivion: The Prospects for Humanity*, 1969*

Written 40 years ago, Bucky Fuller's postulated choice between utopia for all humankind or oblivion for all seemed like a real dilemma at the time. Since then, the world has obviously muddled along between those poles. Now, however, we seem to be approaching a time when it appears that the choice between a more sustainable future and one of great uncertainty, stemming from climate change unprecedented during the Holocene epoch, is more real than ever.

The steady growth of carbon dioxide concentrations in the earth's atmosphere, to levels unseen in hundreds of thousands of years, with attendant global warming, makes the transition to a future based on renewable, non-carbon energy sources ever more urgent. Since residential and nonresidential buildings contribute about 40 percent of the world's carbon dioxide emissions, the focus of carbon reduction policies, programs, and actions must necessarily be on the built environment.

This book is about greening existing buildings, reducing their energy and water "footprints," to more sustainable levels. In a broader sense, this book is about sustainability, but its clear message is that we need to rapidly transform the buildings in which we work, play, live, and learn, as part of the longer-term enterprise of learning to live within the limits of the earth's carrying capacity for human activity.

The Sustainability Revolution

Influenced by Buckminster Fuller's books (and those of many other contemporary visionaries) as a student, I recall many personal discussions in the 1970s about the "carrying

*R. Buckminster Fuller, *Utopia or Oblivion: The Prospects for Humanity*, 1969, Jaime Snyder (ed.), Baden, Switzerland: Lars Mueller Publishers. Reprinted 2008.

capacity" of the environment for human activity and the concern about clean air and clean water that fuelled the first wave of the environmental movement that gave rise to the first Earth Day in 1970. As the air and water began to get cleaner, the first wave of environmental concern subsided and most talk about sustainability vanished during the 1980s.

For about the past 15 years, interest in sustainable buildings and sustainable building operations has been growing in the United States and finding direct expression in both manufacturing and building design, construction, and operations. We might even date this interest to the founding of the U.S. Green Building Council in 1993 by a group of business people and designers influenced by the 1992 U.N. Conference on the Environment and Development and the 1989 formation of the American Institute of Architects "Committee on the Environment."*

Sustainability means many things to many people. What is sustainable building and sustainable enterprise, and how is it connected to the design and operations of buildings and facilities? The most commonly cited definition of sustainability is that of the United Nations' Brundtland Commission on Environment and Development of 1987, which states:

> "Sustainable development is development that meets the needs of the present without compromising the ability of future generations to meet their own needs."†

In a strictly economic sense, sustainability means sustaining economic output. However, most people have adopted a broader view of sustainability, one that recognizes that the essential features of the earth's life-support systems must also be maintained over time.‡ This means that we must learn to live within the flows of sun, wind, and water, as well as biomass growth, and reduce wastes and pollution to levels that can be rendered harmless by natural systems.§

For the past two decades, the corporate social responsibility (CSR) movement has emphasized the social impacts of business practices, with two important elements: acting as a good corporate citizen, attuned to the evolving social concerns of stakeholders, and mitigating existing or anticipated adverse social effects from business activities. Many have seen a focus on CSR as a major strategic business initiative and an essential element in maintaining competitiveness.¶ Combining this concern with the previous two yields the "triple bottom line" approach (Fig. 1.1) that focuses equally on economic, social, and environmental impacts of business practices.

Does sustainability pay? According to a 2008 magazine report, "There is a link between corporate sustainability and strong share price performance. Companies with

---

*David Gottfried, *Greed to Green*, 2004, Berkeley, CA: WorldBuild Publishing, provides a good history of the formation of the U.S. Green Building Council.

†http://www.un.org/documents/ga/res/42/ares42-187.htm, accessed February 8, 2009.

‡John Gowdy, "Terms and Concepts in Ecological Economics," *Wildlife Society Bulletin*, 28(1):26–33, 2000.

§LEED Steering Committee, "Foundations of the Leadership in Energy and Environmental Design Environmental Rating System: A Tool for Market Transformation," Spring 2003.

¶Michael E. Porter, and Kramer, Mark R. "Strategy and Society: The Link Between Competitive Advantage and Corporate Social Responsibility," *Harvard Business Review*, December 2006.

Figure 1.1 **The triple bottom line is a convenient way of thinking about integrating economic, ecological, and social equity concerns into corporate actions.**

the highest share price growth over the past three years paid more attention to sustainability issues, while those with the worst performance tended to do less. Causality is difficult to establish (i.e., does a focus on sustainability lead to more profit, or do profitable companies tend to engage more with sustainability programs?), but the correlation appears clear: the companies that rated their sustainability efforts most highly over this time period saw annual profit increases of 16 percent and share price growth of 45 percent, whereas those that ranked themselves worst reported growth of 7 and 12 percent, respectively."* Most executives in the survey (57 percent) said that the benefits of pursuing sustainable practices outweighed the costs, although 8 out of 10

*The Economist Intelligence Unit, February 2008. Special Report, "Doing Good: Business and the Sustainability Challenge," available at: www.eiu.com/site_info.asp?info_name=corporate_sustainability&page=noads&rf=0, accessed May 11, 2009.

expected any boost to profits to be small. Specifically, sustainable practices can help reduce costs (particularly energy expenditure), open up new markets, and improve the company's reputation.

What are the primary factors driving companies and building owners to incorporate sustainability into their building operations? Three key phenomena come immediately to mind: market transformation, supply push, and demand pull.

MARKET TRANSFORMATION

The initial market transformation for energy-efficient buildings came from the ENERGY STAR program, begun by the federal government in the 1990s. However, by the mid-2000s, one green building rating system had clearly achieved a dominant role in defining what constituted a green building. A program of the U.S. Green Building Council (USGBC), the Leadership in Energy and Environmental Design (LEED) rating system is an important tool for "transforming the building and real estate market towards sustainability while promoting human health, environmental restoration, economic prosperity, social welfare, and equity." By 2007, the LEED rating system had established itself as the primary U.S. green building rating system for nonresidential buildings, rapidly making green building practices and products more commonplace throughout the construction industry.* Despite the 2008-2009 severe economic downturn and widespread concerns about credit availability for building construction, Turner Construction's 2008 "Green Building Barometer" survey found that 83 percent of commercial real estate executives said they would be "extremely" or "very" likely to seek LEED certification for buildings they are planning to build within the next three years.[†] This level of interest in LEED certification is encouraging, representing a necessary but not sufficient condition for the sustainable transformation of the built environment required in the near future.

SUPPLY PUSH

A number of manufacturers are supplying increasing volumes and varieties of products to the sustainable building market. One of the most comprehensive and innovative examples is Interface Global. Interface is the world's largest manufacturer of modular carpet for commercial and residential applications. *Mission Zero* is the company's commitment to completely eliminate any negative impacts the company has on the environment by 2020.[‡] One of the biggest steps Interface has taken in this direction is their service approach to selling carpet. In 1995 Interface began its *Evergreen Lease* program, allowing customers to lease the service of a carpeted space rather than buying carpet. Interface continues to own the carpet, ensuring proper disposal and replacement if a carpet tile becomes stained or the carpeting reaches the end of its useful life.[§]

*See for example my book, *Green Building Through Integrated Design*, 2008, Chapter 2.

[†]http://www.turnerconstruction.com/greenbuildings/content.asp?d=6552, accessed July 22, 2009.

[‡]http://www.interfaceglobal.com/Sustainability/Our-Journey/Vision.aspx, accessed March 3, 2009.

[§]http://www.interfaceglobal.com/Innovations/1995-Evergreen.aspx, accessed January 26, 2009.

When Interface reclaims carpeting, the company separates the carpet fiber from the backing, allowing for a "maximum amount of postconsumer material to be recycled into new products with minimal contamination." This is the first step in Interface's ReEntry 2.0 process.* This process is an example of a "cradle-to-cradle" product cycle, where used or postconsumer materials are perpetually circulated into new products in closed loops. This cycle maximizes the value of the materials and dramatically reduces the volume of waste sent to landfills.†

Through the actions of many manufacturers, green products are increasingly available for use in the building industry, at little or no cost premium compared with conventional or non-green products, in the author's experience.

DEMAND PULL

Some customers are using their sustainability requirements to influence the practices of the manufacturers they purchase from or the businesses they lease space from. For example, suppliers of canned and frozen fruit and vegetables for use by Sysco, a major food service company, are required to submit evidence of integrated pest management (IPM) and sustainable agricultural programs.‡ IPM is an environmentally sensitive approach to pest management that incorporates comprehensive information about the life cycles of pests and includes a range of management options beyond just the use of pesticides.§

In another context, many commercial building owners are opting for LEED certification of both new and existing spaces because they want to be acceptable to the full range of potential tenants. Gary Thomas, sustainability director of CB Richard Ellis, the largest commercial property manager in the United States, explains why many building owners are interested in the LEED for Existing Buildings (LEED-EB) certification:¶

> Primarily, owners in general are motivated by perceived value and market differentiation. There is the perception that a LEED-EB-certified building will maintain a higher occupancy and have higher rental rates than other buildings. There's also the understanding that once buildings start trading again, the sales price of a LEED-EB-certified building will be higher. One of the additional primary motivators is related to tenants. The market is moving in a direction where tenants are starting to include in their requests for proposals (RFPs) for space to lease questions to determine if a building has instituted green practices or if it is LEED-certified. In most major markets, there might be only 5 to 10 percent of the tenants looking for that type of space right now, but as this market evolves and as tenants' education about green building practices and benefits grows, our clients believe that those requests will grow as well. Due to the fact that it takes up to 18 months to complete the LEED-EB certification process, forward-thinking property owners who don't want to be left behind when more tenants start looking for sustainable space to lease are the first to move toward LEED certification.

---

*http://www.interfaceglobal.com/Sustainability/Sustainability-in-Action/Closing-the-Loop.aspx, accessed March 3, 2009.

†William McDonough, and Braungart, Michael. *Cradle to Cradle: Remaking the Way We Make Things*. North Point Press, April 2002.

‡http://www.sysco.com/aboutus/aboutus_pestm.html, accessed March 3, 2009.

§http://www.epa.gov/opp00001/factsheets/ipm.htm, accessed March 3, 2009.

¶Interview with Gary Thomas, May 2009.

GROWTH IN SUSTAINABILITY REPORTING

Sustainability isn't just a response to environmental regulations; it is a framework for approaching the myriad of opportunities available in the current business landscape. As author Marc Epstein explains, "Companies are currently looking to get past the regulatory and compliance aspect of this. The companies that 'get it' are beginning to look at the growth that can come about as a result of sustainability."*

An IBM survey of 250 global business leaders released in 2008 found that businesses have actually assimilated a more strategic view of corporate social responsibility: 68 percent are now utilizing CSR as an opportunity and a platform for growth, through developing new products and services.†

A recent report by the Social Investment Forum (SIF) shows significant growth in sustainability reporting by S&P 100 companies since 2005:‡

- 48 percent increase in the number of S&P 100 companies with sustainability portions of their Web sites
- 26 percent increase in the number of S&P companies producing sustainability reports

A report by KPMG says that more than half the Global 500 companies now routinely issue sustainability reports.§ The question remains open, however, as to how many companies are truly making a transition to more sustainable practices.

Gil Friend, CEO of Natural Logic, a leading sustainability consultant, feels that the economic downturn "is prompting many businesses to re-evaluate their value propositions and priorities. This goes hand in hand with increasing eco-efficiency and creating long-term value through sustainable jobs and energy sources." He also cites several reasons why interest in sustainable practices remains strong despite the 2008-2009 economic recession.¶

- Reducing energy costs from most sustainability initiatives
- Mitigating risk from expected future government controls on carbon emissions
- Taking advantage of opportunities that will likely arise owing to the Obama administration's commitment to green issues
- Securing an attractive return on investment, often a characteristic of well-designed sustainability initiatives

---

*Mark Epstein, 2008, *Making Sustainability Work: Best Practices in Managing and Measuring Corporate Social, Environmental and Economic Impacts*, San Francisco, California: Berrett-Koehler Publishers.

†www-935.ibm.com/services/us/index.wss/ibvstudy/gbs/a1029293, accessed May 11, 2009.

‡*2008 S&P 100 Sustainability Report Comparison*, Sustainable Investment Research Analyst Network, July 2008 (full citation).

§http://www.csrwire.com/press/press_release/13790-KPMG-International-Survey-of-Corporate-Responsibility-Reporting-2008, accessed July 22, 2009.

¶http://featured.matternetwork.com/2009/1/no-downturn-corporate-sustainability-initiatives_8068.cfm, accessed January 23, 2009.

The Global Carbon Problem

Certainly a core issue for building owners is the continuing growth of global carbon emissions from fossil fuel combustion and a myriad of other human activities. Table 1.1 shows a growth in world CO_2 emissions of more than 30 percent between 1990 and 2005. At present, CO_2 concentrations in the atmosphere are about 386 ppm and are growing at the rate of about 3 ppm per year. Most climate scientists forecast significant adverse impacts when levels hit 450 ppm,* at which point average global temperatures would rise more than 4°F, while others are now sounding the alarm that we've already exceeded thresholds for irreversible change and are calling for a reduction in atmospheric concentrations to 350 ppm.† Table 1.1 offers some hope; the world has reduced its CO_2 emissions per unit of gross domestic product (a standard measure of economic output) by 43 percent over a recent 15-year period. As an example, in one Swedish city, Vaxjö, between 1993 and 2008, gross domestic product grew 30 percent while CO_2 emissions fell by 20 percent, through a combination of renewable energy, energy efficiency retrofits, a focus on encouraging bicycle travel and use of public transportation, and many other smaller measures that created a cumulative large effect.‡ (Nevertheless, because of rapid economic growth in various countries, total world carbon emissions still increased by 36 percent over that period.)

TABLE 1.1 ENERGY-RELATED CARBON DIOXIDE EMISSIONS, 1990 AND 2005*

	CO_2 EMISSIONS (BILLION METRIC TONS)		CO_2 EMISSIONS PER CAPITA		CO_2 METRIC TONS PER GDP† (CURRENT DOLLARS, MILLIONS, PPP)‡	
COUNTRY/REGION	**1990**	**2005**	**1990**	**2005**	**1990**	**2005**
United States	4.9	5.8	19.6	19.7	858.0	479.9
China	2.3	5.6	2.0	4.2	1,471.8	592.6
European Union	4.2	4.1	8.7	8.4	626.3	330.1
India	0.6	1.2	0.7	1.1	542.6	336.2
Japan	1.1	1.2	8.9	9.8	469.4	319.4
World	**21.1**	**27.5**	**4.0**	**4.2**	**786.5**	**451.0**

***CO_2 Emissions Data:** Climate Analysis Indicators Tool (CAIT) Version 6.0. (Washington, DC: World Resources Institute, 2009), http://cait.wri.org, accessed July 27, 2009.

†**Gross Domestic Product (GDP) and Population Data:** International Monetary Fund, World Economic Outlook Database, September 2006, http://www.imf.org/external/pubs/ft/weo/2006/02/data/index.aspx, accessed July 27, 2009.

‡PPP represents "purchasing power parity."

*http://www.urbanecology.org.au/news/2007/ieareportstabiliseby2012.html, accessed May 9, 2009.

†www.350.org/about/science, accessed May 9, 2009.

‡www.vaxjo.se/VaxjoTemplates/Public/Pages/Page.aspx?id=1661, accessed May 11, 2009.

TABLE 1.2 GLOBAL ENERGY AND CARBON DIOXIDE EMISSIONS, 2005, AND POSSIBLE SCENARIOS FOR 2050

INDICATOR	2005 (ACTUAL)	2050 BUSINESS AS USUAL (BAU)	2050 STABILIZATION SCENARIO
CO_2 emissions (Gt)†	28	62	14
World population (billions)*	6.5	9.3	9.3
Change in CO_2 emissions	–	+130%	−50%
CO_2 emissions per capita (metric tons)	4.3	6.7	1.5

***Population Data:** U.S. Census Bureau, International Data Base, http://www.census.gov/ipc/www/idb/, accessed on July 28, 2009.

†**CO_2 Data:** © OECD/IEA, 2008, Energy Technology Perspectives: Scenarios & Strategies to 2050, http://www.iea.org/Textbase/techno/etp/ETP_2008_Exec_Sum_English.pdf, accessed on July 28, 2009.

What is clear is that urgent action is needed, just to hold atmospheric CO_2 concentration growth to 450 ppm by 2050. For example, Table 1.2 shows the level of emissions in 2005 and two likely scenarios through 2050, assuming a 37 percent growth in world population.

To meet the demands of the "2050 Stabilization Scenario," carbon emissions would be only 50 percent of 2005 levels, implying a massive reduction in fossil fuel energy use through reliance on renewable energy, energy conservation, and carbon sequestration. To accommodate anticipated world population growth, per capita emissions in 2050 would fall by about 65 percent from 2005 levels.

THE ROLE OF BUILDINGS IN CREATING CARBON EMISSIONS

Many studies have indicated that all forms of buildings, residential and nonresidential, contribute nearly 50 percent of the total carbon emissions in the United States, as shown in Fig. 1.2. This use is split about half between residential uses and half nonresidential, including industrial, commercial, recreational, retail, education, and all the various types of buildings and uses.

We know already that activities in buildings consume nearly 70 percent of all electricity produced in the United States, and that a majority of the electricity is produced through fossil fuel combustion. Hence, buildings directly contribute to a significant amount of greenhouse gas production. In addition, most buildings have boilers, furnaces, and similar devices for direct combustion on-site. Further, buildings indirectly consume energy through workplace commuting (which in fact exceeds direct consumption of energy in offices), building materials such as concrete and steel, office and industrial supplies and materials production, and waste disposal. Add in the indirect energy use in water supply, distribution and treatment, and there is even more energy use directly attributable to the built environment. Table 1.3 shows some of these direct and indirect impacts of buildings.

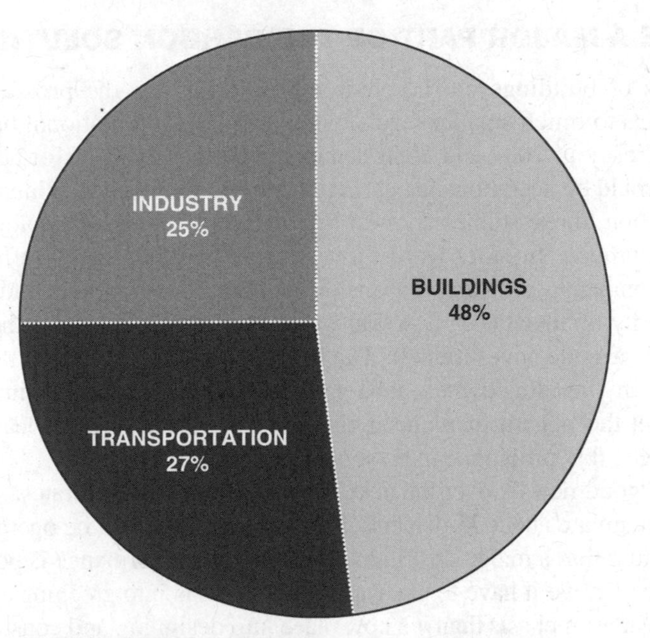

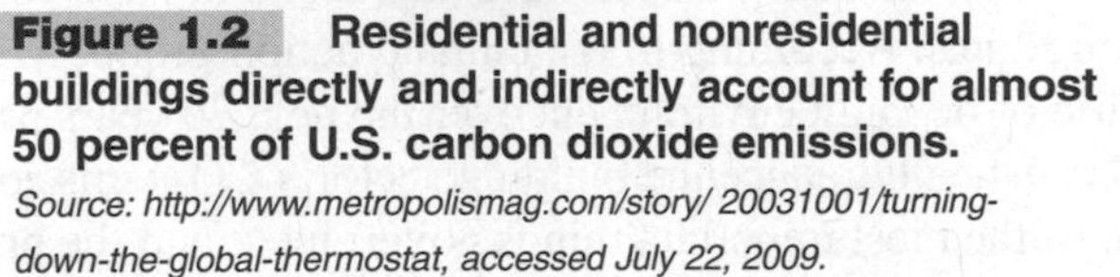
Figure 1.2 **Residential and nonresidential buildings directly and indirectly account for almost 50 percent of U.S. carbon dioxide emissions.**
Source: http://www.metropolismag.com/story/ 20031001/turning-down-the-global-thermostat, accessed July 22, 2009.

TABLE 1.3 ENERGY AND ENVIRONMENTAL IMPACTS OF BUILDINGS IN NORTH AMERICA*

ITEM	PERCENT OF TOTAL U.S. PRODUCTION/ CONSUMPTION	PERCENT OF TOTAL CANADIAN PRODUCTION/ CONSUMPTION
Total energy use	39–40	33
Electricity	68–72	NA
Water (nonindustrial)	12–14	12
Solid waste generation (nonindustrial)	40	25
Carbon dioxide emissions	38	35

*Commission for Environmental Cooperation, 2008, Green Building in North America: Opportunities and Challenges, p. 22. Accessed May 9, 2009, from http://www.usgbc.org/DisplayPage.aspx?CMSPageID=1766. See also: http://www.usgbc.org/DisplayPage.aspx?CMSPageID=1718, Doc. 3340, May 9, 2009.

BUILDINGS ARE A MAJOR PART OF THE CARBON SOLUTION

Given these impacts of buildings on the environment, what are the prospects for reducing these impacts to much smaller levels? Studies by the international business consulting firm McKinsey in 2007 and 2008 demonstrated that the transformation to a low-carbon world could be accomplished at reasonable economic cost, using today's technology.* In addition, these studies showed that *building energy efficiency is the only cost-effective approach*. In other words, investing in building energy efficiency makes money for a company, using any reasonable economic assumptions, indicating the problem is primarily financial (i.e., how can we pay for this upgrade?) rather than economic (i.e., is this a good investment?). The only issue is to overcome the split incentives between many building owners and tenants (i.e., owners make the improvements, but tenants get the benefit of reduced energy costs, the way most leases are structured). Fortunately, this problem can be solved in a variety of ways!

One further bit of "good news": over the next 30 years in the United States, we will collectively build or renovate about 75 percent of all buildings that will be operating in the year 2040, indicating that a major shift in building energy performance is possible. However, for this to occur, we'll have to place a major emphasis into greening existing buildings, an even greater emphasis than we now place into designing and constructing "net-zero-energy" new buildings. According to the climate action group, Architecture 2030, "This transformation of the built environment over the next 30 years provides an historic opportunity to dramatically reduce the building sector's CO_2 emissions."†

For these reasons, one of the most important things government and the private sector can do is to incentivize and finance building energy efficiency upgrades. The Clinton Climate Initiative has promoted this approach since 2007, inducing many of the world's largest banks to make multi-billion-dollar commitments to finance building energy upgrades.‡

As one example, in December 2008, Wells Fargo Bank announced that it had already loaned more than $3 billion for green buildings and renewable energy. Originally, in 2005, Wells Fargo announced a 10-point commitment to integrate environmental responsibility into its operations and business practices. The commitment included a pledge to provide more than $1 billion in lending, investments, and other financial commitments to environmentally beneficial business opportunities within five years. By the end of 2008, Wells Fargo had provided more than $3 billion in financing for green businesses, exceeding its finance goal more than two years early.§

How large is the business opportunity for greening existing buildings? Table 1.4 shows the size and energy use of the building stock in the United States and Canada.

---

*http://www.mckinsey.com/mgi/publications/Investing_Energy_Productivity/, accessed May 9, 2009.

†http://www.architecture2030.org/current_situation/stop_coal.html, accessed May 9, 2009.

‡http://www.greenerbuildings.com/feature/2008/04/22/the-clinton-climate-initiative-a-business-solution-climate-challenges, accessed May 9, 2009. See also: http://www.clintonfoundation.org/what-we-do/clinton-climate-initiative/what-we-ve-accomplished, accessed May 9, 2009.

§https://www.wellsfargo.com/downloads/pdf/about/csr/reports/Three_Year_Progress_Report_FNL.pdf, accessed May 9, 2009.

TABLE 1.4 COMMERCIAL BUILDING OPPORTUNITIES, U.S. AND CANADA

	UNITED STATES*	CANADA
Number of buildings	4.7 million	>440,000
End-use energy consumption, buildings as percentage of total country	17%†	14%
Greenhouse gas (GHG) emissions, buildings as percentage of total country	18% (just carbon)	13%
Growth of GHG emissions, 1990 to 2003	+26% (just carbon)	+44%
Forecasted BAU GHG emissions, 2050 vs. 2005	+28% (2030, just carbon)	+207%

*U.S. numbers from the U.S. Department of Energy Buildings Energy Data Book: http://buildingsdatabook.eren.doe.gov/ChapterView.aspx?chap=3#4, accessed May 8, 2009.
†http://www.architecture2030.org/current_situation/building_sector.html, accessed May 8, 2009.

You can see that there are nearly 5 million buildings ripe for retrofit into energy-efficient structures.*

As we go through our exploration of the existing building retrofit and operations market, you'll see that there are myriad chances to improve the energy and environmental performance of these structures for everyone's benefit. But first, let's take a look at one high-profile project, the Empire State Building in New York City, opened in 1931, the then tallest building in the world. If an 80-year-old building with 300 tenants can make enough changes, in a very tough economic environment, to eventually merit a LEED-EB rating, what's to prevent any other building from claiming the same achievement, with a similar effort?

Case Study: Upgrading and Updating the Empire State Building

Early in 2009, Jones Lang LaSalle and Johnson Controls, along with the building's owner, announced a $20 million energy retrofit for the 2.6-million-square-foot Empire State Building, as part of a larger $500 million building renovation and upgrade

*Canadian numbers from David McLaughlin, CEO, *National Roundtable on the Environment and the Economy*, Ottawa, Canada, presentation at Green Real Estate Conference, Toronto, April 30, 2009.

Figure 1.3 **Once the tallest building in the world, the Empire State Building illustrates many of the special challenges for greening existing buildings.**

program (Fig. 1.3). This project will last until early 2011, at which time the team expects to qualify for a Gold certification under the LEED for Existing Buildings program.*

What was the motivation behind this massive transformation? Dana Schneider, northeast market lead for energy and sustainability services, Jones Lang LaSalle, functioned as the program manager and as the sustainability consultant for the effort. The work that was done on this project was by a team that comprised Jones Lang

---

*April 6, 2009, press release announcing the project, http://216.40.252.67/SocMe/?id=213&pid=197&sid=213&Title=Press+Releases&Template=ContentWithTertiaryNavigation, accessed May 10, 2009.

LaSalle, Rocky Mountain Institute, the Clinton Climate Initiative, and Johnson Controls. The team worked very closely together to develop everything as *a replicable model for building retrofits*.* According to Schneider:

> We evaluated the current status of the building as compared with the [then] current version of LEED for Existing Buildings: Operations and Maintenance and completed an exhaustive feasibility study, credit-by-credit, outlining ways for the building to achieve Certified, Silver, Gold and Platinum status with associated incremental costs and detailed strategies—exactly what they would need to do [to get that rating]. The building has committed to going for LEED-EB Gold.

The building's owner, Anthony Malkin (CEO of Wien & Malkin and representing the Empire State Building Company), was committed to reducing the carbon footprint of this building and all of his properties, which made it a logical decision to pursue LEED certification. However, Schneider says,

> The owner's initial focus was heavily on energy-efficiency, energy use reduction and the resulting reduction in carbon footprint. Subsequently we also discussed the ability for the building to achieve LEED [certification] and the comprehensive sustainability approach associated with LEED in going beyond energy efficiency and carbon footprint reduction. It was not the initial intent; the owner did not go into this with LEED as the goal. His goal was to optimize energy efficiency and reduce the carbon footprint while demonstrating positive economic returns.

Paul Rode is the project manager for Johnson Controls, Inc., the energy service company (ESCO) that implemented the building retrofits at the Empire State Building and is measuring the on-going associated savings. We'll see in this book the important role ESCOs can play in accelerating the pace and scope of building energy retrofits. Rode talks about some of the unique issues of renovating a multitenant building while trying to get major energy savings:†

> The age of the building presented a challenge, because the documentation on how it was built was scarce. We had to do a lot of exploratory work. The building has such a high utilization factor that we didn't have the luxury of vacant floors to look at or machinery that wasn't in use. All of the work had to be done around people without affecting the operations of the place. [From a business perspective, it was really important] to find measures that reduced energy consumption but did not reduce service. We also needed to find a way to get savings out of loads controlled by tenants in such a way that tenants would actively increase their understanding and awareness of consumption. So we had to think of ways to help the tenants reduce their energy consumption.

Rode and his team adopted a whole building technique, which he expects to use on every similar building retrofit in the future.

---

*Interview with Dana Schneider, April 2009.

†Interview with Paul Rode, April 2009.

> It involved modeling the building in a simulation program, so we could understand by changing one system the effect on other systems. Typically in existing buildings, you'll see owners, in one broad brush, apply a particular technology [such as lighting]. It will have an energy-saving effect, but they might be missing many other opportunities by not looking at the building as a whole.

Beyond just analyzing building energy performance, Rode found that it was essential to look at capital investments planned and forecasted in the coming years, to decide how to proceed.

> For instance, an interesting item in this case was that they were looking at increasing air-conditioning loads over the years ahead because they're changing their tenant population from 800 tenants each occupying smaller square footage to 100 or 200 tenants occupying larger spaces. They're also looking at full-floor tenants. Previously, they had two or three per floor, now they're going to have one-third of the building consisting of full-floor tenants. The nuance in that is every square inch of the space is used where previously you had 30 percent of the core area comprising hallways. Now that same core area has workspaces in it and is seeing much higher energy loads. Rather than build a larger chiller plant, we were able to justify major load-reduction measures—insulation and windows—to keep the existing chiller plant at its current size. When considering where the building is expected to be in 10 years, more measures opened up that can be done now.

When all of the analysis was done and the approach for investing in building envelope upgrades and systems was selected, it was then important to bring the building staff up to speed.

> The low-hanging fruit was implemented during the year of the exploratory study. Numerous interviews and seminars were held with the existing building staff, and they implemented schedule changes [right away]. It was customary to keep equipment on longer than it needed to be. Set point changes were also implemented during the study year. The next measures would include reducing the cooling energy. With building controls, it starts with simply turning off systems when not in use or necessary and reducing the demand between the radiator system in the wintertime and the central ventilation system [which of course brings in cold air that has to be heated].

Rode's comments illustrate the importance of the study phase for any major building retrofit, the need for a whole building simulation to investigate proposed measures and the value of having a *bias for action* by implementing easy and understandable improvements right away, without waiting for a study to conclude and a report to be issued. By having Johnson Controls at the table from the beginning, a financial feasibility assessment could be made at each stage and energy savings are guaranteed through the energy services contract.

Let's take a look at the specific measures used in this project and, just as importantly, to the approach used, which will help this type of project be replicated widely in the years ahead. The project consists of eight specific projects, estimated to reduce energy use by up to 38 percent over current (pre-retrofit) levels and to save $4.4 million per year in energy costs, all on an incremental investment of about $20 million. The improvements are expected to place the Empire State Building in the top 10 percent of U.S. commercial office buildings in terms of energy efficiency. The project put particular emphasis on the approach used to generate the retrofit projects.* More than 60 energy-efficiency ideas were considered, before the team settled on the final 17 implementable projects. The projects are grouped into three categories:

Reduce loads

1 Remanufacture more than 6500 windows by inserting a thin film and a gaseous mixture between two existing panes of glass.
2 Adopt daylighting strategies, reduce lighting power density, and use plug-load occupancy sensors to turn off equipment and lights when no one is in a space.
3 Install radiative (infrared reflective) barriers behind more than 6500 radiators to ensure that heat actually goes into the building instead of out through the walls.

Meet remaining loads efficiently

1 Replace more than 300 existing air handling units (AHUs) with new variable air volume (VAV) AHUs.
2 Retrofit existing chiller plant for greater efficiency and control by replacing the "guts" of each unit with new components, while retaining the shell.

Control energy use better

1 Tenant energy management systems, including submeters to allow tenants to access accurate, real-time, and transparent energy use data online.
2 Tenant demand-control ventilation, which involves the installation of CO_2 sensors for control of outside air introduction to chiller water AHUs and direct-expansion (DX) AHUs. One return air CO_2 sensor will be installed per unit in addition to removing the existing outside air damper and replacing it with a new control damper.
3 Improve functioning of direct digital controls (DDC), both new and existing, to optimize heating, ventilating, and air-conditioning (HVAC) system operation and to allow tenant submetering.

None of these approaches is unusual or outside the norm of engineering practice or outside the capabilities of building operators. What is new is the way in which they have been combined to provide an optimum investment for the owner. The project expects to see the following important results:

- Achieve an energy use reduction of 19 percent in the initial phase (the first 18 months until the end of 2010), and gradually increase the savings to 38 percent as the longer-term projects are completed (by the end of 2013).

*http://www.esbsustainability.com/SocMe/?Id=0, accessed May 10, 2009.

- Create a competitive advantage in the marketplace.
- Provide an impetus to increase the number of multitenant building retrofits that seek more dramatic energy use reductions by tackling tenant energy-using systems as well as base-building systems.

What's the larger significance of this project? According to Dana Schneider:

> When the building was built in 1931 it was the tallest building in the world. It was built during the Great Depression. We look at the greening of this building—becoming one of the most energy-efficient buildings in the world—in this economic downturn to be the same beacon that it was when it was built. When it was built it brought hope to millions of people: even in the Great Depression, it was a positive sign of the incredible things the people can achieve. The way we look at it now is that we're doing the same thing. Times are hard and you hear a lot of negative news out there—economically and also with global warming—so look at this icon that is once again setting the standard for what people are capable of achieving in the face of adversity. It's an honor to be part of a project like this.

Summary

Long-term investments in energy efficiency and water savings are the economic drivers for greening existing buildings. However, as we shall see later in this book, for many building owners, there are also substantial economic and financial returns available in improving health and productivity, securing marketing benefits, engaging with the political authorities and developing tools that can be used in a portfolio of building retrofits. In the following chapters, we'll explore how many different types of owners and facility managers are approaching the greening of their buildings and facilities. This effort is vital to reducing the carbon dioxide emissions of the built environment, helping to secure a more liveable future for everyone.

2

THE CHALLENGE OF GREENING EXISTING BUILDINGS

The challenge of greening existing buildings is to demonstrate achievement while still respecting budgets, addressing tenant/occupant resistance to change, and meeting corporate constraints on activities. With the global recession beginning to hit hard on commercial real estate in 2008 and 2009, the challenge of finding investment and debt capital to upgrade existing buildings is significant, even as the returns from such investments continue to increase, something we will describe in far more detail in Chap. 3.

One approach to greening existing buildings is through the adoption of a very specific protocol, either the U.S. Green Building Council's LEED rating system or by securing an ENERGY STAR label for a specific building. Building owners and managers have adopted both approaches, since both provide third-party certification of achievement.

ENERGY STAR assesses buildings according to their relative energy use among similar buildings nationwide, assigning a score based on the percentile ranking and awarding a label only for buildings in the top quartile.* LEED focuses on a broader array of environmental attributes, including considerable focus on energy savings, but also promoting sustainable site selection and land use, water conservation, environmental preferable materials, and waste disposal, along with indoor environmental quality.

We present the LEED for Existing Buildings (LEED-EB) rating system in detail in Chap. 4, but at this point, we will use LEED as a more general surrogate for the growth of interest in greening existing buildings in recent years.

* According to the U.S. EPA, "To qualify for the ENERGY STAR, a building or manufacturing plant must score in the top 25 percent based on EPA's National Energy Performance Rating System. To determine the performance of a facility, EPA compares energy use among other, similar types of facilities on a scale of 1–100; buildings that achieve a score of 75 or higher may be eligible for the ENERGY STAR. The EPA rating system accounts for differences in operating conditions, regional weather data, and other important considerations." http://www.energystar.gov/index.cfm?c=business.bus_bldgs, accessed July 5, 2009.

Growth of the LEED Rating System

The growth of the LEED rating system in the past five years has been spectacular. Beginning in 2005, the number of projects registered to seek LEED certification began to grow rapidly, as shown in Fig. 2.1, rising more than tenfold in just four years. This growth was especially pronounced for new building projects, as represented by the LEED for New Construction (LEED-NC) and LEED for Core and Shell (LEED-CS) programs. As one can infer from Fig. 2.1, on a cumulative year-over-year basis, these programs grew more than 380 percent from the end of 2006 through the end of 2008. This growth took LEED registered projects from a very small percentage to nearly 20 percent market share of all new nonresidential construction in 2008.

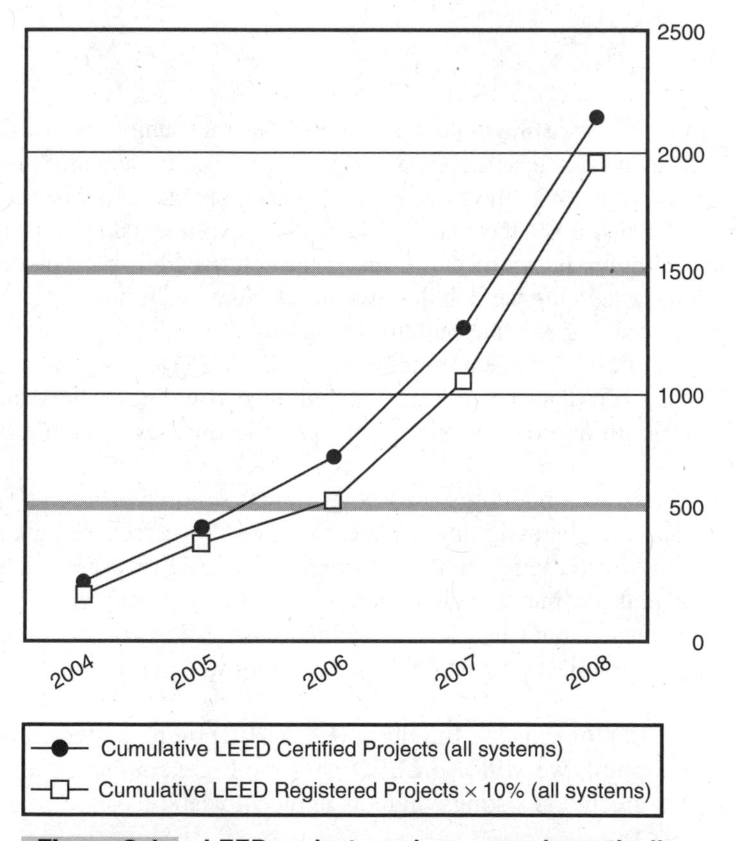

Figure 2.1 **LEED project numbers grew dramatically during the five years, 2004 through 2008, rapidly gaining acceptance as the *de facto* U.S. green building rating system.***

*Note that in Figure 2.1, the number of LEED registered projects is multiplied by 10 percent to put it on the same scale as the number of LEED certified projects.

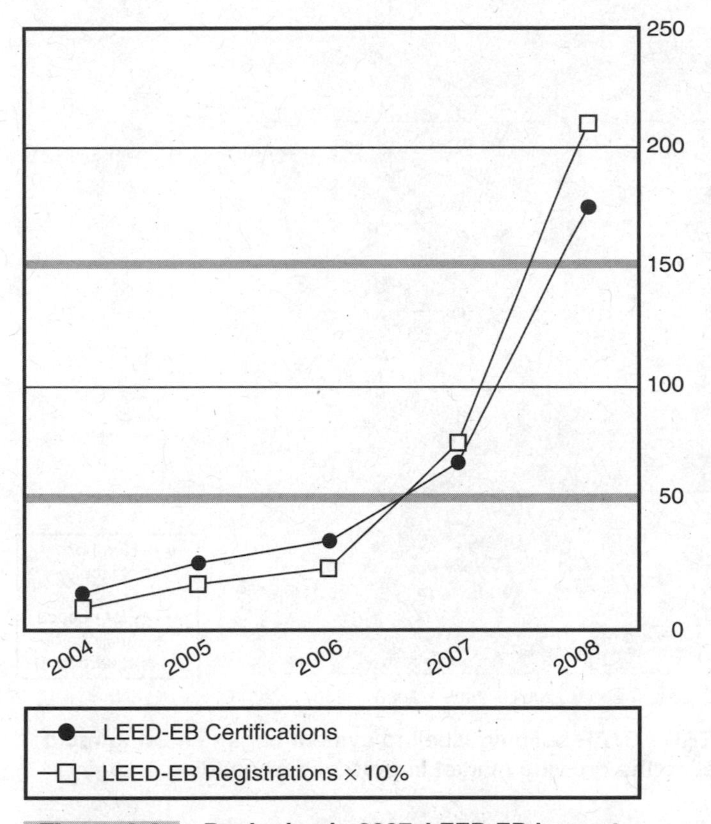

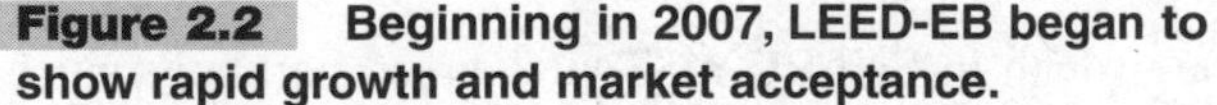

Figure 2.2 **Beginning in 2007, LEED-EB began to show rapid growth and market acceptance.**

Beginning in 2007, however, there was a shift: the fastest growing program began to be LEED-EB, for which new program registrations grew on a cumulative basis by 850 percent from the end of 2006 through the end of 2008, more than double the rate of all LEED programs. This growth is shown in Fig. 2.2. As of July 2009, nearly 3500 projects were registered for eventual certification, while 270 had already been certified.*

Beyond LEED certification, the more established program is ENERGY STAR, which provides a valued label for those buildings in the top quartile of energy-efficiency in operations, for more than a dozen different categories of buildings. ENERGY STAR was created by the U.S. Environmental Protection Agency in the 1990s and has become an accepted mark of quality for building energy performance among commercial building owners, something rare for a government program. However, ENERGY STAR labeling refers only to a building's *relative* performance; in the future, one is likely to see far more emphasis placed on *absolute* energy use, since that is the only means of assessing a building's contribute to CO_2 emissions.

*USGBC data, furnished to the author.

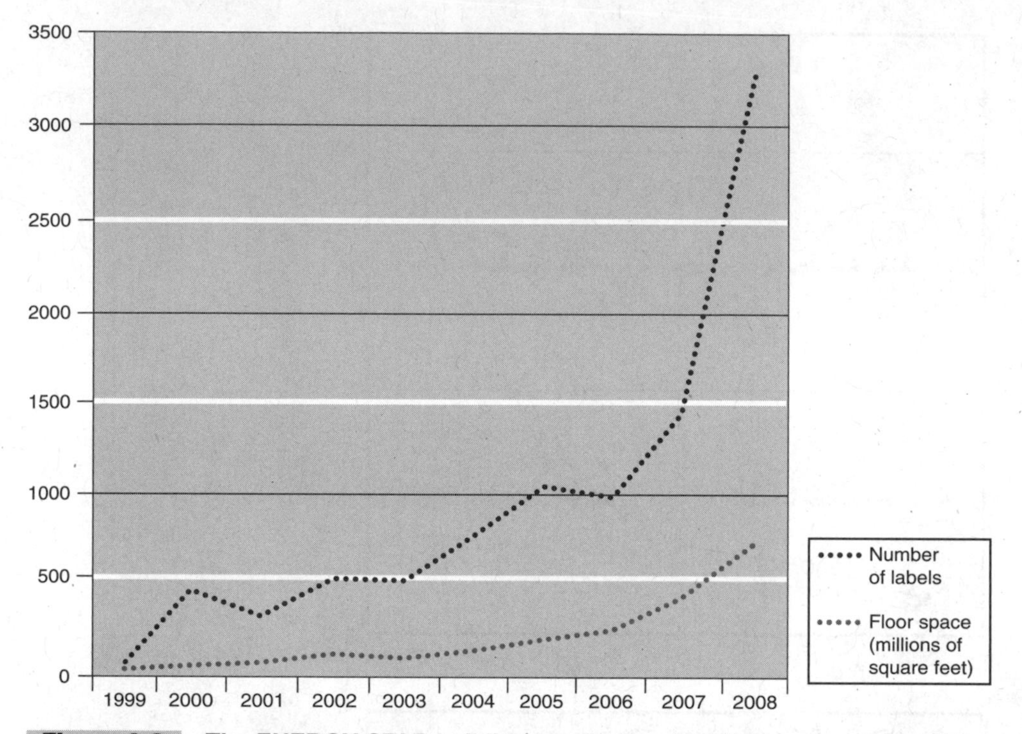

Figure 2.3 **The ENERGY STAR building labeling system began showing rapid growth in 2006, illustrating the growing market importance of building energy efficiency.**

Figure 2.3 shows the growth in ENERGY STAR-labeled buildings since 2004. At the end of 2008, ENERGY STAR had labeled more than 6200 commercial buildings.* As one can see from Fig. 2.3, the program has experienced considerable growth beginning in about 2006, about the same time that the LEED system began to experience rapid growth.†

ENERGY STAR is still a very regional market at present, as this list of the top 10 cities with the number of such buildings in 2008 shows:

- Los Angeles, CA, 262
- San Francisco, CA, 194
- Houston, TX, 145
- Washington, DC, 136
- Dallas-Fort Worth, TX, 126
- Chicago, IL, 125
- Denver, CO, 109
- Minneapolis-St. Paul, MN, 102
- Atlanta, GA, 97
- Seattle, WA, 83

---

*http://www.energystar.gov/index.cfm?c=business.bus_bldgs, accessed May 14, 2009.
†http://www.energystar.gov/index.cfm?c=business.bus_bldgs, accessed May 13, 2009.

Many larger cities such as New York, Boston, Detroit, Philadelphia, Phoenix, and San Antonio are not on this list. The reasons for this are complex, but certainly include the business culture of each city, the local cost of energy and the degree of business benefit seen by a building or facility owner in receiving an ENERGY STAR label.

Drivers to Growth of the Existing Buildings Market

Two key intents of this book are to explore why LEED-EB suddenly became so popular and defining the barriers and incentives to greening existing buildings. In the Introduction, we saw that the concern over energy use and carbon emissions is playing a strong role. But in the case of the Empire State Building, other business factors were also in play: marketing benefit, competitive advantage, an attractive rate of return on discretionary investments, and the availability of financing for the project. Table 2.1 shows some of the key drivers for the growing interest in greening existing

TABLE 2.1 DRIVING FORCES FOR GREENING EXISTING BUILDINGS, U.S. AND CANADA, 2010 TO 2014, RANKED IN ORDER OF IMPORTANCE

		RELATIVE IMPORTANCE	
DRIVING FORCE	COMMENTARY	2010–2011	2012–2014
1. Tenant demand	Tenants are increasingly demanding LEED-certified buildings.	Medium	High
2. Attractive return on investment	Many energy retrofits and LEED-EB certifications are showing high rates of return on investment for owners.	High	High
3. Responsible property investing	Investors and owners committed to corporate social responsibility are asking for LEED buildings.	Low	Medium
4. Future competitiveness	Owners with a longer-term perspective are concerned that their properties' attractiveness might diminish.	Medium	High
5. Stakeholder pressure	Employees, investors, tenants, and communities want green buildings.	Low	Medium
6. Corporate sustainability	Building owners see investing in sustainable measures as an important way to occupy a leadership position.	Low	High
7. Concern about energy prices and future volatility	Energy is the largest cost of building operations and the least controllable. Future price increases could easily outstrip inflation.	Medium	Medium

buildings, as judged by the author. Some drivers are economic, some are environmental, and some are social. The triple bottom line is engaged with both drivers and barriers or inhibitors to greening existing buildings, as ever more businesses seek to create sustainability programs that include building and facility certifications.

TENANT DEMAND

The first driver, tenant demand, has come largely from corporate and governmental lessees, who want to locate in office space that their employees and other stakeholders will find attractive from a green standpoint. Certainly, this has been a factor in the rapid growth of LEED-certified properties among speculative commercial office developers, as evidenced by the rapid growth of the LEED-CS program. For example, cumulative LEED-CS project registrations grew 660 percent from the end of 2006 to the end of 2008, from 325 to 2147. Such projects averaged more than 400,000 square feet (equivalent to a 13- to 20-story office building), and LEED-registered projects represented more than 866 million square feet of space by the end of 2008 and more than $130 billion of new construction value.* This driver should increase to a high level of importance over the next five years.

RETURN ON INVESTMENT

The second driver, an attractive return on investment, is certainly very important. After all, people own buildings to make money. A number of case studies of LEED-EB projects indicate that the "payback" on incremental investment can be as low as two years. What this tells one is that there is a lot of "low-hanging fruit" to be plucked with a systematic look at building operations. For longer payback periods, such as the five-year return forecasted for the Empire State Building, the availability of financing (to pay for the investment) becomes much more critical as a driving factor. This driver should stay at a high level of importance over the next five years. The age of a building is not the sole determinant of whether it should receive an energy performance upgrade; rather one must assess the condition of the building and the return on such investments.

A survey of 750 corporate real estate executives in late 2008, sponsored by Turner Construction Company, found that commercial real estate executives viewed green buildings as having lower energy, operating and lifecycle costs; and higher building values, asking rents and occupancy rates. Respondents also noted that green buildings can generate greater investment returns.† Specifically, 84 percent of executives said that energy costs were lower in green buildings, and 68 percent said overall operating costs were lower. Green buildings create an attractive cost/benefit ratio according to most executives, and are considered to be less expensive than non-green buildings for several key measures of benefits and costs.

---

*USGBC data furnished to the author. At a construction cost of, say, $150 per square foot, 866 million square feet would represent $130 billion of construction value.

†www.turnerconstruction.com/corporate/content.asp?d=6504, accessed May 11, 2009.

RESPONSIBLE PROPERTY INVESTING

The third driver, responsible property investing, stems from the growing power of institutional investors in the real estate investment market, such as public pension funds, labor union pension funds, university endowments, and socially responsible investing (SRI) funds. These funds seek out investments that meet both financial and social objectives and are likely to increasingly favor green investments, both in new and existing facilities. This driver should increase to a medium level of importance over the next five years, from its current low level.

FUTURE COMPETITIVENESS

The fourth driver is future competitiveness. For good reasons, many building owners believe that they will be less competitive in future years without an ENERGY STAR designation or LEED certification. As we will see in the chapter on the business case for greening existing buildings, ever more data are accumulating to support this viewpoint. The owner of the Empire State Building, profiled in Chap. 1, explicitly stated future competitiveness as one of his key reasons for seeking the ENERGY STAR and LEED certifications for the energy upgrade. This driver should increase from medium to a high level of importance over the next five years. (Of course, over time, as more buildings become certified, the competitive advantage of the early adopters will erode and they will have to seek other means to respond to competitive pressures.)

STAKEHOLDER PRESSURE

The fifth driver, stakeholder pressure, can come from a variety of sources, including political sources and internal, in the form of employee pressure. Some cities are now beginning to mandate LEED-EB certification for major retrofits, something the Mayor of Los Angeles announced in April 2009.* Other pressures might come from the need to acquire an ENERGY STAR label or a building energy label (required in California beginning in May 2010).† This driver should increase from a low to a medium level of importance over the next five years.

CORPORATE SUSTAINABILITY

The sixth driver, corporate sustainability, is beginning to emerge as an important impetus to greening corporate real estate. According to KPMG, more than 50 percent of the "Global 500" largest companies already issue sustainability reports. The Global Reporting Initiative listed 1200 reporting companies in 2007. Greening real estate is one positive measure companies can take to implement sustainability policies and to

---

*http://www.ecoseed.org/index.php/general-news/green-politics/green-policies/americas/1453-la-city-buildings-retrofit-to-follow-leed-standards, accessed June 29, 2009.

†http://www.green.ca.gov/NewsandEvents/NewsStories/080424.htm, accessed June 29, 2009.

reduce their carbon footprint.* This driver should increase from a current low level to a high level of importance over the next five years.

CONCERN ABOUT ENERGY PRICES

The seventh driver, energy prices and concern over their volatility, is of medium concern in the short and medium term. In certain parts of the country, the cost of electricity is quite high, especially during summer peak cooling periods, but for most of the country is still reasonably low, with a national average in 2008 of about $0.10 per kilowatt-hour.† However, for most commercial buildings, energy costs represent about one-third of all operating costs and are the least controllable. Therefore, building operators are concerned about how much cost they must pass on to the tenant, or in the case of owner-occupied buildings, how they will budget for future energy prices. Concern over energy price volatility, especially if the United States and Canada severely restrict new coal- or oil-fired power plants, will lead owners to want to "future proof" their buildings over the next five years, with significant energy retrofits.

Case Study: Greening the Chicago Merchandise Mart

Let's take a look at how some of these drivers affected the LEED-EB Silver project for Chicago's Merchandise Mart (Fig. 2.4), the largest multitenant commercial building in the United States at about 4.2 million square feet. The challenge for this project was to engage with the diversity and large number of tenants, as well as to deal with the systems in an older building that had received many upgrades since it opened in 1930. The Merchandise Mart is the world's largest commercial building and largest wholesale design center. The Mart spans two city blocks and rises 25 stories. There are more than three million visitors each year to its retail shops and boutiques, 11 floors of permanent showrooms for furnishings, 10 floors of office space, dozens of trade shows each year, and many special educational, community, and consumer events.‡

The improvements had to be financed within the existing constraints of ongoing maintenance and renovation, and the tenants had to be engaged in the process, at least to assist with the environmentally preferable purchasing requirements for LEED-EB certification. The LEED-EB project began in 2005 and concluded in 2007. Mark Bettin is the vice president for engineering at the Mart.§ He said,

> The LEED-EB certification was more about process than it was about the project. Our goal was to work within the existing confines of the operations and capital budgets. For example, we worked on a number of controls renovations during the LEED perform-

---

*http://www.globalreporting.org/NewsEventsPress/PressResources/Pressrelease_28_Oct_2008.htm and http://www.globalreporting.org/GRIReports/GRIReportsList, accessed June 29, 2009.

†http://www.eia.doe.gov/cneaf/electricity/epm/table5_6_a.html, accessed June 29, 2009.

‡http://www.mmart.com/mmart/about/index.cfm, accessed May 14, 2009.

§Interview with Mark Bettin, April 2009.

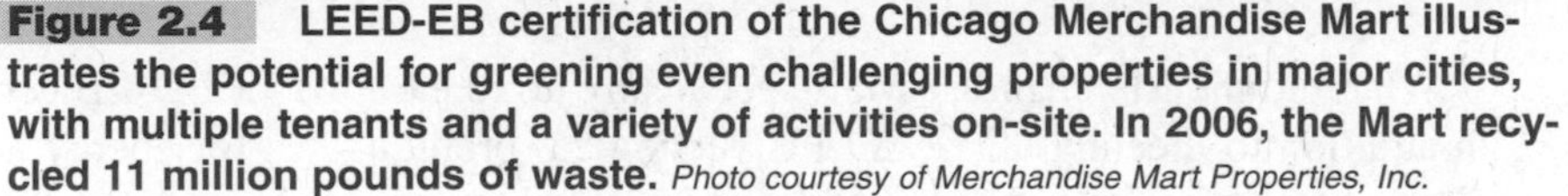

Figure 2.4 LEED-EB certification of the Chicago Merchandise Mart illustrates the potential for greening even challenging properties in major cities, with multiple tenants and a variety of activities on-site. In 2006, the Mart recycled 11 million pounds of waste. *Photo courtesy of Merchandise Mart Properties, Inc.*

> ance period, and we continue to evaluate our controls to this day. In retrospect, we might have reprioritized things. For example, when we went through the evaluation of demand-side ventilation, we made a change to use CO_2 to control heating, ventilating, and air-conditioning (HVAC) systems. As far as the actual breakdown of additional costs for LEED certification, the goal was to work within the existing capital programs within the facility.

At the beginning, the team focused on LEED-EB certification as strictly an operations issue, but later realized that it required a broader approach.

> From early on, we had a lot of supervisors together that oversee housekeeping, engineering, plumbing, and so on, and we went into the certification focused on the things that a typical landlord has control over—the utilities and maintenance that affect the common areas. As we really got into it, we realized the focus is more encompassing because you have to include tenant activities throughout. There was a shift during the project to expand the involvement of others in the company and the project became more encompassing. We started focusing on our meeting-planning group and our tradeshow activities. We did our tenant survey—to get their feedback and buy-in—probably much later than we would have liked to if we were to do it over again.

The Merchandise Mart also recognized during the process that LEED-EB requires a significant effort to engage tenants through communications; it's not just an engineering or facilities management exercise. Everyone has to get involved.

> As a landlord, going through the process became more of an exercise in answering the questions, "How do you get the word out to everybody? How you do the messaging for the tenants, and how do you promote issues or things within the facility that can help the tenants?" Since we're located in the central business district in Chicago, there are a lot of people that use public transportation, and there's also a good-sized segment that rides bicycles. We've had bicycle storage programs here [for some time], but we opened expanded bicycle storage facilities since beginning the LEED-EB process.
>
> Certainly having top-down objectives is important, especially if you are going to spend money in new areas. So support from the top of the company is critical for LEED projects. Originally, the question about achieving certification was posed to (just) us in the operations group, but it's not a question just for operations; the company as a whole was part of the process. It's that kind of all-encompassing objectives that are the key to success. When you have a large company it's always a challenge to get everybody on the same page with the objectives. In that same vein, "green" and "sustainability" are all-encompassing terms and within that everybody has limited resources. So in the end, you need to create goals and objectives that work for your organization and have a decent chance of being attainable. In our case, there are 400 people between operations and property management, so this project was about creating initiatives that involve everybody and that was a big change for us—definitely a positive one.

What did the Merchandise Mart specifically do to earn the LEED-EB certification, in addition to meeting all LEED-EB version 2.0 prerequisites for energy and indoor air quality performance? Some of the notable activities included:*

- In 1990, the Mart began using Green Seal-approved green cleaning products and in 1991 implemented a recycling program, which now includes all forms of paper products, glass, light bulbs, batteries, aluminum, and construction materials.
- In 1996, the Mart became one of the first major property owners in downtown Chicago to use the district cooling system now known as Thermal Chicago, thus helping to reduce the discharge of ozone-damaging chlorofluorocarbons (CFCs).
- In 2006, the Mart joined Clean Air Counts, a voluntary initiative to reduce smog-forming pollutants and energy consumption in the greater Chicago area. Part of the campaign strategies included utilizing only low-volatile organic compound (VOC) cleaning products, paints, and building materials, as well as energy-efficient lighting and alternative workplace transportation options. To date, the Mart has reduced pollution by 264,018 pounds, the largest reduction by a commercial building.
- Also in 2006, the Mart's recycling program saved over 13,000 trees, and water conservation efforts saved 5.5 million gallons of water.
- The Mart and 350 W. Mart Center, also owned and managed by Merchandise Mart Properties, Inc. (MMPI), recycled nearly 11 million pounds of waste in 2006.

---

*www.delta-institute.org/news/Mart_Journey_Release.pdf, accessed May 11, 2009.

- In 2007, the Mart began using recycled paper for all business purposes, installed motion-activated lighting sensors in restrooms, and began using lower-wattage lighting fixtures wherever possible, made an I-Go hybrid car available to tenants and employees 24 hours a day, and retrofitted exit lights to require less energy.

You can see that one of the big challenges in organizing a LEED-EB project at a very large organization such as the Merchandise Mart is just getting the knowledge and buy-in of the entire company, not just the operations department. This takes a lot of upfront communications and a sincere desire to get everyone's cooperation. It can't just be an upper management "top-down" initiative. This point is further discussed in Table 2.2, which deals with the barriers to greening existing buildings.

TABLE 2.2 GREEN BUILDING RETROFIT MARKET—MARKET BARRIERS/INHIBITORS RANKED IN ORDER OF IMPACT (U.S. AND CANADA), 2010–2014

INHIBITING FORCE OR BARRIER	COMMENTARY	RELATIVE IMPORTANCE	
		2010–2011	2012–2014
1. Divergence between capital outlays and operating budgets	Many organizations tend to skimp on capital outlays that will improve operating results, even putting them in different budget categories, especially in the public sector.	High	Medium
2. Split incentives between tenants and landlords	Triple net leases prevalent in the United States and Canada reduce landlord incentive to invest in energy savings that will benefit tenants.	High	Medium
3. Perceived costs far outweigh benefits	Perceptions from early green buildings of significantly higher costs vs. tangible benefits.	Medium	Low
4. Benefits of energy investments not proven	Even if tenants want to reduce energy costs, there's little proof that base building measures will do that.	Medium	Low
5. Incentives are not strong enough to change behavior	Energy prices are still low and tax and other political incentives are not significant enough to change behavior across the nation.	High	Medium
6. Lack of debt financing for energy upgrades	Most building owners must borrow money to finance energy upgrades that have more than a one-year payback.	High	Medium
7. Energy costs vary widely across the United States and Canada	Energy prices vary by a factor of two to four, depending on location, making it hard to incentivize and justify on a national basis.	Medium	Medium
8. Organizational dynamics	Getting buy-in from everyone in the organization is often difficult.	Medium	Low

We've seen examples of the positive driving forces for greening existing buildings. What about inhibiting factors or barriers that are preventing widespread adoption of this approach? These must surely exist, since we are at the very beginning stages of green building operations and maintenance activity. The following section deals with these issues.

Barriers to Greening Existing Buildings

In Table 2.2, you'll see some of the key barriers to greening existing buildings in the United States and Canada, which I have ranked in order of market impact over the next two years, and also out to five years. These barriers suggest actions that can be taken by both government and the private sector to remove or reduce the effect of these inhibiting factors on the rate of green building renovations and operations.

These barriers or inhibiting factors reduce the growth rate of green building renovations and make them more costly. In a 2008 survey cited above, 750 corporate real estate executives rated the following as presenting an extremely or very significant obstacle to green construction: higher construction costs (61 percent), the length of the payback period (57 percent), and the difficulty quantifying the benefits of green building (43 percent).*

DIVERGENCE BETWEEN CAPITAL AND OPERATING COSTS

The first barrier is the divergence between capital and operating budgets in most private sector and public organizations, which makes it difficult to secure funds for investments in energy-efficiency measures that have more than a one-year payback, in terms of savings versus investment costs. Obviously, this is a situation where financing becomes critical, or else a situation that requires the organization to take a longer-term perspective on energy conservation. This barrier can be overcome by having a clear policy for energy efficiency investments and a clear path for acquiring the necessary financing. For example, some larger building owners may make a commitment to achieving ENERGY STAR labels for most of their properties and create "revolving funds" which take the savings from energy-efficiency investments and use that as a source of capital for the next round of investments.

SPLIT INCENTIVES BETWEEN TENANTS AND OWNERS

The second barrier, split incentives between tenants and building owners, affects only rental properties and only in the short run. In other words, landlords with triple net leases just pass along the energy costs to tenants and don't have a direct incentive to make energy efficiency investments that will benefit primarily the tenant (assuming that the landlord pays a percentage of common-area charges). However, landlords are

---

*www.turnerconstruction.com/corporate/content.asp?d=6504, accessed May 11, 2009.

continually renovating and improving their buildings to keep tenants happy. In this respect, it's in a landlord's self-interest to make these improvements and to LEED certify a building, to keep tenants over the long run.

PERCEIVED HIGH COSTS OF GREENING

The third barrier, the perceived high costs of greening an existing building, in comparison with the future benefits, is probably lower now than it has been as people get more familiar with and more comfortable with green building retrofits. Nonetheless, each project must be "cost justified" by showing benefits well in excess of costs.

UNPROVEN FUTURE BENEFITS

The fourth barrier, the unproven nature of future benefits, is easier to overcome with whole building energy modeling, something that is affordable for large buildings (and which was employed in the Empire State Building example in the previous chapter). In addition, most people already believe that lighting and building controls retrofits are cost-effective, when coupled with educating facilities staff. However, a number of studies have shown a wide variability in building energy performance, even compared with modeled projections of savings, largely based on occupant behavior, so some skepticism about future energy savings is certainly warranted.*

INCENTIVES TOO SMALL TO CHANGE BEHAVIOR

The fifth barrier stems from incentives that are insufficient to change behavior. There is a federal tax deduction in place through the end of 2013, of up to $1.80 per square foot, for building energy efficiency retrofits that achieve 50 percent savings against a referenced standard (ASHRAE 90.1-2001). Partial deductions of up to $0.60 per square foot can be taken for measures affecting any one of three building systems: the building envelope, lighting, or heating and cooling systems,† but this level of savings may be quite difficult to achieve in retrofits. Other incentives might include utility rebates for conservation investments. Of course, for buildings owned by public agencies or nonprofits, tax incentives are of little value.‡

LACK OF FINANCING FOR ENERGY UPGRADES

The sixth barrier is lack of financing for energy upgrades. A 2009 survey by Johnson Controls found that access to capital is constraining many business leaders from

*Cathy Turner, and Frankel, Mark, 2008, *Energy Performance of LEED for New Construction Buildings*, Final Report, March 4, 2008, New Buildings Institute, White Salmon, WA, available at www.newbuildings.org/research.htm, accessed May 11, 2009.

†www.energystar.gov/index.cfm?c=products.pr_tax_credits, accessed May 11, 2009.

‡Incentive programs for energy efficiency can be found at the Database of State Incentives, www.dsireusa.org, accessed May 11, 2009.

making such investments, with 42 percent of more than 1400 respondents citing this barrier.* Certainly the worldwide credit crunch in 2008 and 2009 has made borrowing money, even for clearly beneficial reasons, much more difficult.

WIDE VARIABILITY OF ENERGY COSTS

The seventh barrier concerns the wide variability of energy costs in various regions of the United States and Canada, making it difficult for national firms to put company-wide policies in place when energy costs might easily vary by a factor of two or three between locations. The same holds true for the state and local incentives, which vary considerably among regions and even neighboring utility districts.

ORGANIZATIONAL DYNAMICS

Finally, there is the barrier of organizational dynamics. In a multitenant building, it takes participation from nearly all the tenants to achieve a LEED-EB rating, and that can be very difficult to achieve, as we saw in the Merchandise Mart case study. Even in a single-tenant building, getting everyone to adopt policies, provide data, and cooperate during the "performance period" for the project can often be quite difficult.

Project Profile: PepsiCo Corporate Plaza, Chicago

In December 2008, PepsiCo announced that its downtown corporate plaza in Chicago received a LEED-EB Silver certification (Fig. 2.5). In achieving this distinction, the project reduced energy use by 10 percent in less than a year, cut water use to 37 percent below current building code performance standards, and eliminated almost 226 metric tons of greenhouse gas emissions through energy-saving programs. "Our LEED certification is a tremendous accomplishment and a testament to the passion of our employees who created a 'Green Team' devoted to making environmental responsibility an integral part of our corporate culture," said Jim Lynch, a senior vice president with the company.†

The LEED certification is part of a larger commitment by PepsiCo to cut energy and water use. In 2009, the U.S. Environmental Protection Agency (EPA) awarded PepsiCo a 2009 ENERGY STAR Sustained Excellence Award in recognition of its

*Johnson Controls, "2009 Energy Efficiency Indicator Report," May 6, 2009, available at www.johnsoncontrols.com/publish/us/en/news.html, accessed May 11, 2009.

†www.allenmatkins.com/emails/nltr-green/gbu-dec22.htm, accessed May 11, 2009.

Figure 2.5 **In achieving Silver LEED certification, the PepsiCo building in Chicago cut energy use 10 percent and reduced water use 37 percent below current codes.** *Photo by Bruce Critelli, courtesy of PepsiCo.*

continued leadership in energy efficiency.* In 2008 alone, PepsiCo reduced energy costs by $90 million. PepsiCo's mission to reduce energy consumption is part of the company's commitment to sustainable growth, defined as Performance with Purpose. PepsiCo has announced goals to reduce water consumption by 20 percent, reduce electricity consumption by 20 percent, and reduce fuels consumption by 25 percent per unit of production by 2015, compared to 2006 consumption levels.

---

*www.metrogreenbusiness.com/news/green.php?title=epa_recognizes_pepsico_with_2009_energy_&more=1&c=1&tb=1&pb=1, accessed March 8, 2009.

First Blue, Then Green

Both the drivers and barriers to greening existing buildings deal significantly with money. In today's constrained economic environment, one that is likely to be with us for several years to come, getting the resources to make the investments necessary to achieve an ENERGY STAR label or LEED-EB certification can be challenging.

One lesson we can draw from the considerations in this chapter and the case studies of the Empire State Building and PepsiCo is that the most successful LEED-EB projects always begin with a focus on reducing energy consumption, with its attendant economic return, and then proceed on to securing other environmental and health benefits. Since the ENERGY STAR logo color is blue, some have called this strategy, "first blue, then green." It also makes sense because energy savings are likely to provide the most immediate and tangible benefits of a green building "extreme makeover."

Texas-based consultant Mychele Lord, formerly employed with a major developer, concurs.

> The utility [energy savings] line item pays for LEED certification. A lot of these green operational practices have become low-to-no-cost. To obtain the LEED certification, the cost to do that is recaptured through the utility line item, mainly energy. As energy costs continue to rise, the paybacks will become shorter. At most of the buildings that I see, the payback is within the first year. *

Gary Thomas of CB Richard Ellis has a similar take on the situation, expecting energy savings to remain the major justification for LEED-EB upgrades:†

> The major component of LEED-EB that will provide a payback is going through the retrocommissioning process or having an ASHRAE Level 2 Audit completed and instituting the measures that are recommended. The reduced energy consumption and subsequent cost reductions are where you'll derive your savings and experience a payback. There are few measures that take place as part of the LEED-EB certification that will generate a direct financial return. Water efficiency is another possibility, but the benefit depends greatly on the sewer and water fees in a specific market. Even though potable water is becoming more scarce, it is still entirely too inexpensive. In the Atlanta market, we had a 1.5-million-square-foot three-building project where they replaced water fixtures to meet the LEED guidelines and there was a significant return because of the water usage fees in that market. But for the most part, if a client is primarily focused on the financial return, they need to have the retrocommissioning work completed.

However, Thomas does see, at least in some leading markets, a more holistic perspective on the part of building owners, including an assessment that the entire market is moving toward LEED certification as a basic standard for top-of-the-line office

---

*Interview with Mychele Lord, LORD Green Strategies, Inc., Dallas, May 2009.

†Interview with Gary Thomas, May 2009.

buildings. As he noted (Chap. 1), "Primarily, owners are motivated by perceived value," and the perception is that tenants are beginning to move in this direction.

So we've returned full circle, to our first market driver in Table 2.1, tenant demand. Nothing will move the commercial market faster to the green existing building stock than a clearly perceived market demand by significant tenants that want LEED-certified space. For governmental, education, nonprofit, and other non-commercial buildings, the driver is often public policy or organizational policy. For example, more than 900 U.S. mayors have committed their cities to reducing greenhouse gas emissions by 2012 to below 1990 levels.* That will be an impossible goal to meet, without attending to the energy use of existing buildings. Similarly, more than 600 college and university presidents have signed a similar pledge.† More than 20 states have adopted similar policies for new construction and major retrofits, including the State of California under the urging of Governor Arnold Schwarzenegger.

Theresa Townsend is an architect with the California Department of General Services, who successfully guided the team that achieved the state's first LEED-EB Platinum rating, for the California Department of Education building. As a result of the project, the team created an online tool kit to allow other state agencies to approach a LEED-EB project efficiently. Townsend says:‡

> For all of the credits that were harder, we responded with a [specific] tool in the tool kit. There's a presentation online that talks about some of the sustainable activities that the state does, so that when other state agencies do LEED-EB they can refer to it for help in getting started. In the tool kit, we have several things including the Best Practice Manual: Better Building Management for a Better Tomorrow, which is a tool that anybody can use worldwide. It's meant to help everyone put together a manual for their own buildings. It's based on how we manage our buildings, but it's not organized according to LEED-EB, credit-by-credit; [instead] it's meant to be a tool for the custodians and building engineers to follow. All of the requirements for LEED-EB are embedded in it. There are at least 12 credits in there, and all of the prerequisites are addressed. So it's a great tool for people to use. We also have a chart with module tasks one through 10. These 10 tasks will help you put together a best practice manual along with surveying your building and collecting purchasing data for LEED-EB credits. There is also a task included to help you improve your energy efficiency with retrofits.

While relatively few public agencies are yet using LEED-EB as an organizing principle for operations and maintenance, the State of California's experience as an innovator has lightened the burden for both public and private organizations wishing to emulate its approach.

---

*http://usmayors.org/climateprotection/agreement.htm, accessed July 21, 2009.

†http://www.presidentsclimatecommitment.org, accessed July 21, 2009.

‡Interview with Theresa Townsend, March 2009. Tool kit is available online at: http://www.green.ca.gov/GreenBuildings/EBtoolkit.htm, accessed May 11, 2009.

Summary

The importance of greening existing buildings lies in two key facts: existing buildings constitute the vast majority of the energy-using building stock at any given time, and they have huge impacts not only on energy and water use, but on the lives of people who occupy them. Because the United States' economy is so dynamic and the country's population continues to grow, there is great turnover in the building stock, which provides an opportunity to change things for the better, within the space of one generation. Commenting on this situation in 2005, one authority noted the following statistics for the United States:*

- The total building stock equals approximately 300 billion square feet.
- Every year, we tear down approximately 1.75 billion square feet of buildings.
- Every year, we renovate approximately 5 billion square feet.
- Every year, we build new approximately 5 billion square feet.
- By the year 2035, approximately three-quarters of the built environment will be either new or renovated.

Because of these facts, everyone in the building construction and operations industry has an opportunity to influence the future energy use and carbon emissions from this sector. While the main focus of most projects is on improving building energy performance, it's also vital that improvement measures for existing buildings move beyond solely energy considerations to a full range of environmental upgrades. The green building market has been growing largely through new construction, but since 2007, the fastest growing segment has been in the LEED-EB program, which is explained in Chap. 4.

*http://www.architecture2030.org/current_situation/hist_opportunity.html, accessed May 14, 2009.

3

MARKETS FOR GREENING EXISTING BUILDINGS

We've seen the importance of greening existing buildings and have had an introduction to the barriers and incentives for doing it. Now the question is: who's going to do it and how does the market reward greening of existing buildings? This chapter focuses on the tangible and sometimes intangible benefits of greening private-sector buildings. The easiest place to start is with buildings that have been labeled ENERGY STAR or certified under the U.S. Green Building Council's (USGBC's) LEED system. Both are national in scope, cover most building types, and are widely accepted as authentic, third-party-verified measures of building energy and environmental performance.

For commercial building owners, the key economic factors for justifying green and energy-efficient building investments are increases in:

1. Average rents
2. Average occupancy rate
3. Resale value

Over time, these three factors, along with others such as tenant retention, ease of getting financing, lower insurance costs, etc., will help overcome any adverse consequences from the costs of greening buildings and certifying them. What are the facts today?

Commercial Benefits of Green Buildings

In the past two years, several important studies of the commercial benefits of green buildings all pointed in the same direction: green buildings make more money for their owners.

THE COSTAR STUDY

The first major study of the benefits of ENERGY STAR labeling and LEED certification was performed by Professor Norman Miller of the University of San Diego, California, using a database from a commercial property database leader, CoStar. Published in March 2008, and based on data from 2004 through 2007, the study produced the conclusions shown in Table 3.1. Subsequent studies by other research teams confirmed the results of this study.

The CoStar/Miller study analyzed more than 1300 LEED and ENERGY STAR buildings, representing about 351 million square feet in CoStar's commercial property database of roughly 44 billion square feet, and assessed those buildings against non-green properties with similar size, location, class, tenancy, and year-built characteristics to generate the results. The study analyzed only multitenanted, Class A office buildings of 200,000 square feet or more, with at least five stories and built since 1970.

At this time, it's certainly clear that one factor accounting for the "green" premiums might be the small supply of green buildings, which account for just a fraction of the total U.S. building stock (less than 1 percent of total leasable space in CoStar's database). The study suggests that while the number of LEED-certified and ENERGY STAR-rated buildings continues to grow, the supply has not kept pace with demand, hence the premiums.

It's obvious that the gains in average rents and average occupancy are enough to pay a considerable amount for green building upgrades, while still retaining a healthy profit on operations and adding to building value in the case of resale.

THE RICS STUDY

Using a similar database of 893 properties, compared with 10,000 noncertified buildings within a quarter-mile, a team of researchers at the University of Maastricht in the Netherlands and the University of California, Berkeley, produced a study for the U.K.'s Royal Institution of Chartered Surveyors (RICS) in 2008 that showed certified green buildings had 6 percent higher rents and achieved a 16 percent greater sales price (based on 199 sales).*

TABLE 3.1 ECONOMIC BENEFITS OF ENERGY STAR AND LEED BUILDINGS

	ENERGY STAR	LEED CERTIFIED
No. of buildings	960	360
Rent increase	$2.40/square foot	$11.33/square foot
Occupancy increase (nominal percentage)	3.6%	4.1%
Resale value increase	$6.20/square foot	$171/square foot

http://www.costar.com/josre/doesGreenPayOff.htm, accessed July 24, 2009.

*Piet Eichholtz, Kok, Nils, and Quigley, John, "Doing Well By Doing Good? An Analysis Of The Financial Performance Of Green Office Buildings in the USA," *RICS Research Report*, March 2009, London.

Using a database of almost 500 buildings for which energy use data were available, the analysis also showed that a $1 savings in energy costs from increased thermal efficiency yields a return of roughly $18 in the increased valuation of an ENERGY STAR-labeled building.

The study indicated that the average effective rent for the 7488 control buildings in the sample of rental office buildings is $23.51 per square foot. On the other hand, at the average size of these buildings, the estimated annual rent *increment* for a green building is approximately $329,000. At prevailing capitalization rates of 6 percent, the incremental value of a green building was estimated to be about $5.5 million more than the value of a comparable unrated building. (At a capitalization rate of 10 percent, the increase would still be $3.3 million.) Given that most of the LEED-EB certifications we will see in this book cost less than $500,000, you can see that the return on investment (ROI) on this incremental capital outlay would be quite large.

RESPONSIBLE PROPERTY INVESTING STUDY

A major international movement has grown up around the notion of responsible property investing (RPI). In 2009 a study using the RPI approach showed that ENERGY STAR-labeled buildings commanded 5.9 percent higher income per square foot and exhibited a 13.5 percent greater market value.* These benefits derived from 10 percent lower utility costs, 4.8 percent higher rents, and 1 percent higher occupancy rates. These buildings also sold at 0.5 percent lower cap rates than non-labeled properties, providing a higher multiple of net annual income in determining building value. In all other respects, ENERGY STAR buildings showed appreciation and total returns similar to other office properties.

In addition, other green building attributes such as location close to transit and in urban regeneration areas also showed strong positive results. Properties near transit in the suburbs had 12.7 percent higher net incomes, 16.2 percent higher market values, 0.3 percent lower cap rates, 1.1 percent higher annual appreciation, and 0.9 percent higher annual total returns than other suburban office properties. Properties near transit in central business districts (CBDs) had 4.5 percent higher net incomes, 10.4 percent higher market values, and 0.2 percent lower cap rates (but their appreciation and total returns were similar to other CBD office buildings). Properties in or near urban regeneration areas in CBDs had 2.4 percent lower net incomes, consistent with their economically distressed locations, but they still had 1.1 percent higher values per square foot, 0.5 percent lower cap rates, and appreciation and total returns at par with other CBD office properties.

*Gary Pivo and Fisher Jeffrey, 2009, "Investment Returns From Responsible Property Investments, Energy Efficient, Transit-oriented and Urban Regeneration Office Properties in the US from 1998-2008," Working Paper, Responsible Property Investing Center, Boston College and University of Arizona, Benecki Center for Real Estate Studies, Indiana University, accessed May 14, 2009, from www.u.arizona.edu/~gpivo/.

THE RREEF STUDY

The RPI approach has become a worldwide movement. A 2008 review of the benefits of green buildings by RREEF, a unit of Deutsche Bank, found that:*

- Real estate developers and managers are adopting greener business practices in all regions of the world, driven by the favorable financial returns for greener buildings, owing to increasing energy costs and the significant savings in building operations resulting from well-executed green designs or renovations.
- Globalization is reinforcing and accelerating sustainable property development and operating trends throughout the world. Multinational corporations and global investment firms are especially important in establishing greener real estate business practices worldwide through their criteria for securing tenant space and making investments.
- Major corporate tenants are seeking greener facilities in order to attract and retain workers, differentiate their products, improve their image to consumers, and satisfy shareholder demands, all of which have ties to environmental concerns. Such firms often set minimum global energy-efficiency and/or green standards for the buildings they lease or buy.
- Property investors who want to diversify their portfolios and leverage their expertise are looking globally for their acquisitions and developments, further spreading sustainability practices into new regions.
- Greener business practices are also being driven by increasingly prescriptive government regulations. In a growing number of countries, developers must build sustainably, owing to greener building codes and green building certification requirements for planning approvals.
- Global international environmental and green building organizations, investor pressure groups, and multilateral institutions also are playing key roles in driving greener building standards.

Market Trends

RREEF research reported in February 2009 its expectation that "major real estate markets—the markets where institutional investors focus their attention—will be pushed even faster to the tipping point where green building becomes the market standard." The research predicts that older, less-efficient conventional buildings will actually have their market value discounted in the years ahead.† Even with the

---

*www.kennedyusa.com/PDFs/RREEF_Global_Greening_Trends_2008.pdf, accessed May 14, 2009. Original study, by Andrew Nelson, 2008, "Globalization and Global Trends in Green Real Estate Investment," RREEF Research, September, pp. ii-iii.

†Andrew Nelson, "How Green A Recession?—Sustainability Prospects In The U.S. Real Estate Industry," RREEF Research, Paper No. 70, February 2009, p. 8. Available at https://www.rreef.com/cps/rde/xchg/ai_en/hs.xsl/3157.html, accessed July 25, 2009.

continuing global economic recession, government policies will continue to accelerate the push toward greener buildings, as will tenant demand, especially from corporate real estate executives. There is also "no pronounced indication that major institutions are pulling back from their greening commitments" as investors.* In this context, greening existing buildings, especially upgrading energy efficiency, can be seen as a "defensive strategy," since these less-efficient properties risk "market decay" in the form of lower rents and higher vacancies, "as tenants increasingly migrate to more modern, greener buildings."

THE ROLE OF REGULATORS, TENANTS, AND INVESTORS

Scott Muldavin is executive director of the Green Building Finance Consortium. Here's his take on the situation:†

> The great thing about today—and over the past 18 months—is there's been a dramatic shift in the importance of green or sustainable buildings, particularly energy efficiency for the regulatory community, the tenant/occupier community, and the investor community. When you look at the value or financial performance of any green existing building investment, those three groups (regulators, tenants, investors) really drive the value, so *that increase in the level of importance today has overcome one of the most important obstacles*.

Muldavin believes that green and energy efficiency upgrades to buildings may also benefit from a concern that investors have about the risk of functional obsolescence in the buildings they might buy.‡

> Many of the top investors in this country are already developing acquisition screens for any existing building that they might buy. They want to know the potential functional obsolescence in a building—the way that you can make it energy efficient or sustainable. *The key issue—and this is really an appraisal term—is the cost to cure. You don't want to buy a building today that the cost to cure the obsolescence, which would be lack of energy efficiency or sustainability, is so high that it doesn't make any sense.* You don't want to own buildings today that are exposed to all of the risk. Functional obsolescence can lead to economic obsolescence because in the future you might have tenants, regulators, and investors that don't want that building.

---

*Ibid., p. 12.

†Interview with Scott Muldavin, April 2009.

‡According to Scott Muldavin, "The key to supporting sustainable property financial decisions is to 'prove-up' the level of increased regulator, space-user, and investor demand at the property level. Each property decision will have a unique set of regulator, space-user, and investor criteria and issues that need to be evaluated in the context of the factual context of the property and market conditions. Capital providers and appraisers need property specific analysis, with implications on rents, occupancies, tenant retention, expenses, and risk clearly articulated in order to fully credit a property with 'value' benefits beyond cost savings." Personal communication, June 2009.

ENERGY EFFICIENCY FINANCING DISTRICTS

Lisa Galley is head of Galley Eco Capital in San Francisco. She points out an entirely new source of capital for green building upgrades that's developing very rapidly in California and is likely to spread around the country. She says:*

> In terms of financing, first of all you have to define the financing. You have to ask, "Is there specific financing that's exclusively for green buildings?" That is not coming from the big national banking market. All of the money and forces of capital are coming from the public sector in a big way. For example, here in California, we have a large movement towards financing programs that are getting started from cities and counties called Energy Efficiency Financing Districts. Basically cities are putting out bonds and using these bonds to provide loans to commercial property owners that they can use to retrofit and upgrade their buildings.
>
> For example [in April] the County of Sonoma [in northern California] announced a $100 million program exclusively for the Energy Efficiency Financing District program. They're the first county government to do so. All of the cities in the county said that they want to offer these loans to their residential and commercial property owners. San Francisco has it in the works. Berkeley has done it. The City of Palm Desert has done it. There are eight or nine cities in California that have done this, in addition to the County of Sonoma. This is very new. All of this activity has happened in the past 18 months.

In spite of the commercial building recession, there are strong forces propelling the greening of commercial buildings at a rapid pace. These forces create opportunity for building owners to create additional value through green building retrofits and renovations, a movement that is likely to accelerate during 2009 and continue through the following five years at least.

Marketing Benefits of Greening Existing Office Buildings

If green buildings really deliver short-term marketing benefits, we should be able to find some good examples in a number of cities that illustrate the results of the studies cited above. Since most of the studies cited are based on new buildings, it's instructive for the purposes of this book to try to find LEED-EB commercial office projects that illustrate the same benefits. Here's one such project.

CASE STUDY: 100 PINE STREET, SAN FRANCISCO

Let's take a look at one typical greening effort, the 100 Pine Street building in San Francisco. The 35-story, 441,000-square-foot, multitenant building is the first such office structure in California to receive a LEED-EB certification, a distinction reached in

*Interview with Lisa Galley, April 2009.

Figure 3.1 **The 35-story office tower at 100 Pine Street in San Francisco is the first such building in California to receive LEED-EB certification, which should provide over time significant marketing benefits.** *Courtesy of Unico Properties.*

mid-2008 (Fig. 3.1). The building also has an ENERGY STAR label; it is unique from both a building management and leasing perspective, according to the owner.* To achieve the LEED-EB certification, the project team invested $1,292,302 (net after utility rebates) to save nearly 800,000 kilowatthours per year of electricity, with a current benefit of $180,660, about a 14 percent annual return on investment. A $10,000 investment in low-flow water fixtures expects to save 1.5 million gallons per year, with an annual utility savings exceeding $11,000. The energy upgrade included adding direct digital controls (DDC) to two levels, converting to a variable air volume (VAV) system from a constant

*Personal communication, Sharon Mead, Unico Properties, June 2009.

air volume system, installing variable frequency drives on all motors, installing heat and motion occupancy sensors in hallways, stairways and other common spaces, and retrofitting garage lighting with T8 lamps. A comprehensive recycling program dating from 1998 saves about $36,000 per year in avoided disposal costs, diverting 75 percent of building waste from landfill, at a cost of about $5500.

Wesley Powell of Jones Lang LaSalle is the leasing agent for 100 Pine, managed by Unico Properties LLC. He says:*

> As a leasing guy, my job is to lease the building and to keep it well leased at the best possible rent for the landlord. Just a few years ago, the LEED effort probably wasn't really translatable into deals or increased rents, but it just made people feel good. That's changed. We all knew that it was a good thing for the building, but it wasn't until just recently that sustainability has been brought to top of mind. The CEOs of top-performing corporations are making sustainability a goal; and that is working its way into office lease decisions. LEED, Green Globes or similar "green" certification has quickly become something that can be translated into more opportunity for the building from a leasing and marketing perspective. At 100 Pine, in 2009, we now see the benefits and results of ownership's decision to become green.

From a tenant's perspective, particularly in a city like San Francisco, where green living is quite accepted, Powell is seeing more tenants put green features on their "must have" checklist for office space:

> LEED certification, more importantly, just sustainability, doesn't have to cost a tenant or building owner more money. In fact, long-term, it saves money. What we're seeing from tenants is that even with the economic downturn, where one would think tenants would decide against green decisions like choosing a sustainable, "green" carpet (one that costs an extra five percent); we're still seeing tenants continue forward with their sustainable commitments. That's a good thing to see for the right reason; our environment. I'd say that in 2007 I saw 0 percent of tenants put sustainability on a checklist as a nice-to-have item. In 2008 I saw 5 to ten percent put it on their checklist as a nice-to-have item. In 2009, we're seeing 20 to 30 percent and I think it's going to be increasing very quickly. Soon, it will be a must-have item. One of the major reasons is green carpet, green paint, sustainable construction is quickly becoming commonplace so it's not costing more or much more. I lease 12 high-rise buildings in downtown San Francisco, and each owner is making the commitment to sustainability. If you are not green, you're going to be behind the eight ball, going to lose deals, be priced lower, and ultimately lose asset value.

Greening the Hotel Industry

Hotel operators are in the unique position of seeing direct bottom line impacts from energy and water savings, reduced waste disposal costs, and possibly reduced health

*Interview with Wesley Powell, April 2009.

claims from using more environmentally friendly pest management practices and green cleaning practices. In addition, there are direct marketing benefits for being a "green hotel." Little wonder that green hotels are one of the fastest growing market segments of the industry.

Greg Reitz is a green hotel consultant who helped with the certification for the first LEED-EB certified hotel, the Ambrose, in Santa Monica, California (Fig. 3.2). For him, the case for greening a hotel is clear:

> Hotel operators have long known the value of incorporating green practices into their operations. As more hotels begin to make claims about their green policies, however,

Figure 3.2 The Ambrose Hotel in Santa Monica, California, was the first LEED-EB certified hotel in the country. *Photo courtesy of the Ambrose Hotel.*

some hotel operators have begun to seek certification to add a level of credibility only possible through a reputable third-party verifier. While programs specialized for hotel operations such as Green Seal have the benefit of being customized for hotels, the LEED-EB certification has an advantage of being internationally recognized across all industries and buildings types.*

CASE STUDY: THE ORCHARD HOTEL, SAN FRANCISCO

The first hotel in San Francisco to be LEED-EB certified was the Orchard Hotel near Union Square (Fig. 3.3). Stefan Mühle is the hotel's general manager. He spoke about how the efficiency process developed, initially through a need to cut operating costs after the 9/11 attacks and the subsequent recession that dramatically reduced business and personal travel.†

> Opening in the year 2000, Orchard Hotel was not built as a green hotel. It was just built as a new, conventional-type of hotel property. After 9/11, we really looked into opportunities to reduce expenses. We started chipping away on that by having each department head look into his or her domain and come up with a few creative solutions on how we could save money. The housekeeping department, for example, discovered cleaning products that are natural, less abrasive to the guest rooms, better for the environment and not any more expensive. In fact, if applied properly they would be less expensive. In maintenance, we found a rebate program with Pacific Gas and Electric (PG&E) that offered free light bulbs, if you discarded your old incandescent lamps and replaced them with compact fluorescents. Every department participated in this and slowly but surely, we started to go green. We didn't give it a name back then. That's the paradox really because there are so many people out there that say going green is going to cost you money. When in fact, for five or six years we were just trying to save money.

This is a key point. For the hotel industry, cost cutting can lead naturally to green solutions. Since it's hard to raise rates, the best way to increase profit is to cut costs, but it must be done without harming the guest experience. Working with partners such as electric and water utilities was very important for the hotel.

The direct impetus for the LEED-EB certification was a new hotel, the Orchard Garden Hotel, built nearby in 2006 for the same owner. This hotel achieved LEED for New Construction basic certification, the first for a hotel in California. Then, according to Mühle, the owner said, "Now it really does make sense to get both hotels on the same pedestal and make sure that they all have the same types of certifications." In looking around for opportunities, the company came across LEED-EB as a way to certify the existing property. Mühle says:

---

*www.greenlodgingnews.com/Content.aspx?id=3006, accessed May 14, 2009.

†Interview with Stefan Mühle, May 2009.

Figure 3.3 **The Orchard Hotel was upgraded to become a green hotel using the LEED-EB system, largely through a series of continuous improvements made over a period of years.**
Photo courtesy of James Yudelson.

The owner of the hotel, Mrs. Huang, is 85 years old and she is 100 percent behind this. She had numerous family members—her father, husband and daughter—all pass away from cancer or cancer-related illnesses within a relatively short amount of time. She said, "If I can build a new property or even if I can retrofit an old property and make it a healthier environment for its occupants, I definitely would like to do so."

Sometimes the champion of the idea can come from anywhere. While greening an existing hotel obviously requires an effort by a lot of people, including vendors, employees, management, and ownership, each successful project does require one person to say, "let's do it." In this case, it was the personal interest of the owner in creating a healthier property for guests and workers.

Greening Existing Retail Buildings

Many large retail store chains have begun to build new LEED-certified stores in the United States and Canada, as well as in Europe (with the U.K.'s BREEAM certification system or others that are evolving in such places as France and Germany), but greening existing buildings has barely started.* Many types of stores constitute the retail sector, including clothing, grocery, restaurants, and the entire gamut of shopping, entertainment, and eating destinations. So far, without strong consumer demand, the push to green existing stores has been basically nonexistent.

However, one store type that lends itself well to LEED-EB certification is the grocery store, for several reasons. Food stores use a lot of energy: think of 24/7 refrigeration and all the energy for cooking and washing in the prepared foods department. Grocery stores also use a lot of water, and they occupy a considerable site area. They also have large waste disposal costs. Finally, they tend to be large chains with centralized purchasing, so that many of the LEED-EB programs can be easily implemented.

CASE STUDY: STOP & SHOP

A major grocery chain, Stop & Shop, located primarily in the eastern United States, has implemented the LEED-EB system in more than 50 stores.† Stop & Shop's parent company, Ahold, has a strong corporate responsibility commitment based on a partnership with customers to build a more sustainable future. Ahold operates 1300 stores along the East Coast, including the Stop & Shop chain.

In 1998, Stop & Shop developed what they called the Low Energy SuperStore (LESS) prototype.‡ As a result, Stop & Shop/Ahold set a goal of building a superstore that uses about one-third less electricity than conventional supermarkets. To target transformative changes, the company focused on savings in lighting and heating, ventilating, and air-conditioning (HVAC); super-efficient refrigeration, systems integration; and building envelope improvements. In 2001, they piloted related innovations by opening a LESS facility in Foxboro, Massachusetts. The value of the model is demonstrated by annual electricity savings of 8 million kilowatthours, which eliminates emissions of nearly 1000 tons of CO_2 annually.

A few years later, the company decided to benchmark its latest store prototype, in Southbury, Connecticut. Store 621 was an ENERGY STAR-labeled model that opened

---

*See Jerry Yudelson's forthcoming 2009 book, *Sustainable Retail Development: New Success Strategies* (Springer) for a fuller description of green building in the retail sector.

†Interview with Leo Pierre Roy, Vanasse Hangen Brustlin, Inc. (VHB), November 2008.

‡Case Study—Stop & Shop. Retrieved December 28, 2009, from www.cleanair-coolplanet.org/information/pdf/StopShop.pdf.

in 2005. Stop & Shop stores have excellent energy efficiency—a company review confirmed that stores built by Stop & Shop after the LESS facility were more sustainable, considering particularly their energy use.

In mid-2007, Stop & Shop began the USGBC's Volume Certification program, using the LEED-EB program as the basis for store certification assessments. The 51 Stop & Shop grocery stores in the certified portfolio are a subset of a much larger group of company stores that share many similar characteristics, making them excellent candidates for the volume LEED-EB certification process. All of the buildings are built from a common specification; further selection criteria included preliminary LEED-EB checklist evaluations, ENERGY STAR ratings, store management/ownership, location, and age. All of the selected stores are located in or near New England.

In May 2008, after about an year's effort, the project team succeeded in achieving LEED certified-level status for the 51-store portfolio, representing nearly 3.4 million-square-feet of retail space. Stop & Shop is the first company and first supermarket chain in the United States to be awarded LEED-EB certification in this manner.

The business case for Ahold/Stop & Shop The most prominent factors in making a business case for LEED were the ability to use the system as a framework for creating new design metrics and the benefit of reduced certification costs per store. The switch from single-building certifications to a volume perspective with attractive economies of scale is critical to giving larger retailers cost-effective incentives to comprehensively address their environmental impacts. From a marketing perspective, LEED is an internationally known standard, which appealed to Stop & Shop as a nationally distributed retailer with considerable brand equity.

USGBC's Volume Certification program helped Stop & Shop to further standardize environmentally responsible programs in their stores by integrating green operations into multiple existing buildings in their portfolio all at once, using the LEED-EB rating system. The certification process met Stop & Shop's overarching goal: to confirm through third-party validation that it was successfully applying sustainable principles to store operations.

What did Stop & Shop do for LEED-EB certification? Energy and water savings were critical elements in both the economics and environmental footprint of the stores. To achieve energy efficiency gains in existing stores, Stop & Shop used cool, white reflective roof membranes, reducing solar heat gain and therefore lowering the demand for air conditioning, and also added extra layers of insulation to hold heat in during the winter.

For energy savings, Stop & Shop focused particularly on product lighting. By specifying more efficient lighting and mechanical systems that produce less waste heat, Stop & Stop saves a great deal of electricity. The stores further conserve energy through appliances like ultraefficient refrigeration and HVAC units. Stop & Shop's advanced refrigeration designs more accurately match the specific refrigeration needs of products in different display cases while at the same time minimizing energy consumption. In addition, waste heat from refrigeration units is used to preheat water for in-store use and to provide space heating.

Summary

Before turning to specific programs for upgrading buildings, it's important to understand the market dynamics of greening existing buildings. From a macroeconomic perspective, energy efficiency upgrades represent the most cost-effective way to meet growing energy demands. From a microeconomic perspective, recent studies have shown that energy-efficient and certified green buildings merit higher market values, greater rents, and higher occupancies. From a corporate sustainability viewpoint, greening existing buildings is a direct way to reduce a company's carbon footprint. As a result, corporate real estate managers in the United States have begun to decide in favor of greening both owned and leased buildings, seeing many economic benefits from this switch. Green buildings offer many marketing benefits for building owners and tenants, including opportunities for creating new green "brands" and also "future-proofing" their real estate against both future energy price increases and also value erosion as the trend toward green buildings continues to grow. Marketing benefits will vary by geographic location, building and tenant type, and other factors, but they are present in all privately owned real estate.

4

UNDERSTANDING GREEN BUILDING RATING SYSTEMS

Anyone who left the United States and Canada four years ago (say in 2005) for a long trek into the wilderness of some other continent could be forgiven for expressing astonishment at how far the movement toward green and sustainable building design, construction, renovation, development, operations, and maintenance has progressed in that period. Just consider some of the following events:

- Nearly 1200 companies worldwide issued some form of sustainability report in 2007, according to the Global Reporting Initiative, including half of the Fortune 500 companies.*
- In 2008, the mega-retailer Wal-Mart announced a $500 million commitment to energy-efficiency in its operations, then went on to become the largest seller of compact fluorescent light bulbs, effectively lowering their price by more than 50 percent overnight. Wal-Mart's new high-efficiency HE.5 store prototype for the western United States expects to save up to 45 percent of energy use of a baseline superstore.†
- In late 2008, Wells Fargo Bank announced that its green business lending for LEED-certified buildings, energy efficiency, renewable energy, and corporate sustainability lending had surpassed $3 billion since 2005.‡
- Oil futures surged past $140 per barrel during 2007 and 2008, triple the level of just three to four years earlier, before falling back to the $50 to $60 level in the spring of 2009, owing in part to reduced demand from the global recession.§

*Global Reporting Initiative, www.globalreporting.org, accessed May 17, 2009.
†http://walmartstores.com/FactsNews/NewsRoom/8163.aspx, accessed May 22, 2009.
‡https://www.wellsfargo.com/downloads/pdf/about/csr/reports/Three_Year_Progress_Report_FNL.pdf, accessed May 22, 2009.
§http://www.wtrg.com/daily/crudeoilprice.html, accessed May 22, 2009.

- By early 2009, more than 900 U.S. mayors, in all 50 states, had signed the Mayors' Climate Change Initiative (many more to follow), committing to reduce CO_2 emissions 7 percent below 1990 levels by 2012.*
- At the same time, more than 630 college and university presidents had signed onto the American College and University Presidents Climate Commitment to initiate the development of a comprehensive plan to achieve climate neutrality on their campuses as soon as possible. While they develop their plans, the campuses also committed to take two or more significant tangible actions to reduce greenhouse gases until they develop a more comprehensive plan.†
- The British Government's 2006 "Stern Review" estimated the likely impact of global climate change would be to reduce world gross domestic product (GDP) by 20 percent at the end of the present century, unless actions are taken in the next decade to reduce the growth of CO_2 emissions. The report made several other dramatic predictions: Unabated climate change could cost the world at least 5 percent of GDP each year; if more dramatic predictions come to pass, the cost could be more than 20 percent of GDP. However, the cost of reducing emissions could be limited to around 1 percent of global GDP. Each metric ton of CO_2 emitted causes damages estimated at $85, but emissions could be cut at a cost of less than $25 per ton. Shifting onto a low-carbon path could eventually benefit the world economy by $2.5 trillion a year.‡
- In March 2009, President Barack Obama made green jobs a centerpiece of his economic recovery program, launching nearly $100 billion worth of energy-efficiency upgrades in public buildings.§
- The LEED for New Construction rating and certification system received project registrations from about 20 percent of the new construction market in 2008, while showing growth above 50 percent annually for the four years ending in 2008.¶

Global Warming and Climate Change Concerns Spur Existing Building Upgrades

As mentioned in Chap. 1, this backdrop of news and reports, along with changing political and economic circumstances, has increased the pressure on building owners and managers to take action to reduce energy use in buildings and to adopt green building measures.

Building energy conservation is increasingly recognized as a major way to attack CO_2 emissions linked to climate change. The 2007 study by McKinsey showed that

*U.S. Conference of Mayors, www.usmayors.org/climateprotection/newsroom.asp, accessed May 17, 2009.
†www.presidentsclimatecommitment.org/html/press.php, accessed May 17, 2009.
‡http://www.guardian.co.uk/politics/2006/oct/30/economy.uk, accessed May 22, 2009.
§http://www.whitehouse.gov/issues/energy_and_environment/, accessed May 22, 2009.
¶Derived from USGBC data supplied to the author, May 2009.

simple technologies such as lighting energy use, better building glazing and insulation, and more efficient heating, ventilating, and air-conditioning (HVAC) systems were the only measures that could significantly reduce carbon dioxide emissions and still represent a net economic gain for society.*

Other studies show that commercial buildings are directly responsible for about 18 to 20 percent of all U.S. CO_2 emissions, primarily from electricity generation to support their operations. For property development and management companies who take their sustainability commitment seriously, building energy conservation has become an important factor to consider in planning future building upgrades, operations, and maintenance programs.

Rick Pospisil is director of facilities for USAA Realty Company's FBI facility in Chicago, which is part of the national portfolio of approximately 65 million square feet. He says:†

> USAA Real Estate Company's approach to sustainability efforts is to drive the business case. Specifically, the team effort to improve efficiency and reduce the environmental impact was driven by the desire to maximize occupant comfort as well as improve financial and environmental performance. It makes intuitive sense that if waste is eliminated and the building operates more efficiently, it should cost less to operate. But we also know that by becoming more efficient and eliminating excess chemicals, we can better control temperature, improve indoor air quality and thus make our tenants more comfortable.

For the 11-story FBI building in Chicago, managed by USAA Realty, the company managed to secure a LEED-EB Platinum rating (Fig. 4.1). This building was a "build to suit" originally completed in 2006. Assessing the costs of LEED certification, Pospisil says:

> The majority of the costs incurred to achieve certification were associated with the consultant and certification fees. Other changes made within the building, that would have been done not to achieve LEED certification but to increase efficiency and reduce costs, realized a net savings of $0.25 per square foot. In short, we saved more than we spent, including costs for certification fees, making changes (like adding faucet aerators, changing irrigation/landscaping etc.), and consulting costs. Taking this a step further, a $0.25/square foot net savings, with an 8% cap rate, creates a potential positive impact on the asset value of $1,568,225 with an 833% internal rate of return and a net present value of $485,855.

Recognizing this high net payoff from LEED certification, Pospisil says that USAA Realty plans to proceed with LEED-EBOM certification of all of its properties:

---

*"A Cost Curve for Greenhouse Gas Reduction," *McKinsey Quarterly*, 2007, vol. 1, www.mckinseyquarterly.com/article_page.aspx?ar=1911&L2=3, accessed May 27, 2007.

†Interview with Rick Pospisil, via e-mail, April 2009.

Figure 4.1 Continuous improvement measures in Chicago's FBI building increased its ENERGY STAR score from 78 to 95. *Photo courtesy of USAA Real Estate Company.*

> This high level of certification further solidifies that our efforts and commitment to greening is an expansion of focus on overall efficiency and that the LEED programs are consistent with our overall commitment not only to energy efficiency but also to environmental stewardship and sustainable operations. This year (2009) we created a majority of the national policies that go beyond energy and into overall environmental sustainability and completed significant upgrades to our preventative maintenance platform. We will put our entire portfolio of office properties through the LEED-EB certification process and hope to have all of them certified soon.

This attitude is typical of many commercial building owners, for example, Jim Rock, senior vice president for leasing at Unico Properties LLC, Seattle, says:

> We've pushed hard to retool our company to be greener and more efficient. We've adopted a green philosophy: Unico believes it is in the best interest of our company, our communities and future generations to protect the environment and conserve natural resources. Unico is benchmarking all properties through the ENERGY STAR program and are seeing results in properties that were built without green principles in mind. Now, 10 of our 13 office properties have gained an ENERGY STAR rating.*

*Interview with Jim Rock, via e-mail, May 27, 2007.

CASE STUDY: U.S. BANCORP TOWER, PORTLAND, OREGON

For the existing 42-story U.S. Bancorp Tower in Portland, Oregon, Unico chose to use LEED-EB certification (Fig. 4.2); the company says, "LEED is the best recognized accreditation in the market that represents Unico's commitment to sustainable practices, energy conservation and innovative ideas. We also wanted to 'green up' our building and find ways to make it more efficient. LEED gave us rigorous parameters and a template to document our efforts."*

Unico began serious capital investments to lower energy consumption in the building prior to applying for LEED-EB, including adding variable-frequency drives (VFDs) to every air handler and pump in the building from 2003 to 2006. Given the total dollars involved ($1.8 million), the company says it had the good fortune of

Figure 4.2 Known locally as "Big Pink," the U.S. Bancorp Tower is Portland's largest office tower; built in 1983, the building provided considerable challenges for the pending LEED-EB certification. *Photo courtesy of Unico Properties LLC.*

*Interview with Margot Crossman, via e-mail, April 2009.

partnering with owners who understood that conservation was not only the right thing to do, but that it made financial sense as well—they secured the capital to complete the projects by working with the Energy Trust of Oregon and used the State's Business Energy Tax Credits to reduce net capital outlay.* Electrical energy savings were estimated at six million kilowatthours per year, worth about $450,000 at today's energy prices.

Unico also had to replace every single plumbing fixture in the building with low-flow equivalents, again at significant capital cost, because simply by virtue of the age of the building (25 years old), nothing met the LEED minimum requirements for water efficiency. The company says, "The water conservation really surprised us—the low flow fixture change had a three-year payback at current water/sewer rates. We would not have investigated the plumbing efficiency if it weren't for the LEED-EB requirement. Frankly, it was not something we expected to have such a quick return on investment, but happily it was not only the right thing to do, but also the fiscally responsible thing to do."

Finally, in terms of the cost of acquiring the LEED certification, beyond using outside consultants to complete much of the documentation, Unico relied on their vendor partners to help out. For example, when the building purchased new high-efficiency particulate air (HEPA) filter vacuum cleaners for the janitorial staff to meet one of the LEED requirements for green cleaning, Unico asked the janitorial service to provide a comprehensive audit of all their chemicals, verify the compliance, and gather the documentation, all of which was done at no additional direct cost to the project.

The Universe of Green Building Rating and Certification Programs

There are two major certification programs in the United States for building operations: LEED and ENERGY STAR.† Additionally, the Building Owners and Managers Association International (BOMA) Canada offers the *Building Environmental Standards* (BESt) recognition and certification program. The Canada Green Building Council also expects to begin offering a Canada LEED-EB program beginning in 2009 that will be modeled on the U.S. Green Building Council's (USGBC's) LEED for Existing Buildings Operations and Maintenance (LEED-EBOM) rating system.‡

---

*Ibid.

†A third rating system, Green Globes, is used primarily in the residential sector in the United States and has almost no acceptance or use in the commercial sector. In September 2006, the U.S. General Services Administration, the country's largest landlord, reviewed five green building rating systems (including Green Globes) and reported to Congress that it would use only LEED to evaluate its future projects.

‡www.cagbc.org/leed/systems/existing_buildings/index.php, accessed May 17, 2009.

The ENERGY STAR Label

Introduced in 1992, ENERGY STAR is a federal program administered by the U.S. Environmental Protection Agency (EPA), jointly with the U.S. Department of Energy. An ENERGY STAR building is in the top 25 percent of energy efficiency of all buildings of its type, denoted by an ENERGY STAR rating of 75 or higher (on a scale of 1 to 100). ENERGY STAR is based on actual energy usage (usually calculated as equivalent BTU's* per square foot per year) for a given building type, adjusted for climate zone and building type. In this respect, it differs significantly from LEED, because LEED certifies buildings partly based on projected future energy use, using a model that calculates energy use reduction from a "code" or "baseline" building of the same size at the same location. ENERGY STAR ratings are available for 15 diverse building types, including commercial offices, corporate offices, government offices, hotels and motels, healthcare, higher education buildings, K-12 schools, and retail and grocery stores. Table 4.1 shows the number of ENERGY STAR labels in each building type. More than 300 project case studies are available for commercial offices with ENERGY STAR certification.† Appendix II explains the ENERGY STAR program in more detail.

In 2006, the CoStar Group, the country's largest provider of commercial real estate information, added the ENERGY STAR rating to its national database of more than two million commercial properties.‡ Having a searchable database of ENERGY STAR-rated buildings for commercial tenants, property owners, and investors makes the rating even more valuable, because companies looking for real estate can search specifically for energy-efficient buildings. Additionally, a 2008 survey for the EPA revealed that nearly 75 percent of the U.S. public is aware of the ENERGY STAR rating system.§

ENERGY STAR labeling has grown dramatically in popularity over the last decade, but especially since 2004. By the end of 2008, ENERGY STAR had certified more than 6200 buildings for their "best in class" energy efficiency. These buildings represent more than 800 million square feet and save an estimated $1.5 billion annually in lower energy bills. ENERGY STAR certified about 1700 new buildings in 2008.

Many major commercial real estate developers use both ENERGY STAR and LEED to enhance the value of their properties. For example, Hines is a major commercial development firm based in Houston and a strong proponent of both the ENERGY STAR rating and the LEED certification systems. With offices in more than 100 cities in 16 countries, and controlled assets valued at about $25.6 billion, Hines is one of the largest sustainable real estate organizations in the world. As a company priority, Hines has labeled more than 130 buildings, representing approximately 76 million square feet in the ENERGY STAR program. In addition, Hines has certified, precertified, or registered under the various LEED programs more than 98 projects, representing approximately 65 million square feet.¶

*One kilowatthour (thermal) is equivalent to 3,414 BTUs.

†www.energystar.gov/index.cfm?c=comm_real_estate.bus_comm_realestate, accessed May 17, 2009.

‡www.costar.com/josre, accessed May 17, 2009, "Does Green Really Pay Off?"

§http://www.energystar.gov/index.cfm?c=news.nr_news, accessed May 17, 2009.

¶Hines Green Office press release, May 2009.

TABLE 4.1 ENERGY STAR-LABELED BUILDINGS BY CATEGORY (AS OF 2008 YEAR-END)*

BUILDING TYPE	NUMBER OF LABELED BUILDINGS	PERCENT OF TOTAL LABELED BUILDINGS
Office	2,509	40.1
Supermarket	1,461	23.4
K-12 school	1,424	22.8
Hotel	369	5.9
Retail	183	2.9
Hospital	80	1.3
Bank/financial institution	74	1.2
Courthouse	37	0.6
Warehouse	26	—
Residence hall	23	—
Medical office	19	—
Cement plant	19	—
Auto assembly plant	15	—
Above and other	13	1.8
Total	**6,253**	**100.0**

*http://www.energystar.gov/index.cfm?c=business.bus_bldgs, accessed May 13, 2009.

As of early 2009, ENERGY STAR building labels comprised about 2500 office buildings, 1460 supermarkets, 1420 K-12 schools, and 370 hotels. More than 340 banks, financial centers, hospitals, courthouses, warehouses, dormitories, and—for the first time—big-box retail buildings have also earned the ENERGY STAR label. More than 35 manufacturing plants such as cement, auto assembly, corn refining, and—for the first time—petroleum refining were also recognized. According to the EPA, buildings that earn an ENERGY STAR rating use about 40 percent less energy than average comparable buildings.*

In early 2009, the Canadian national government agency Natural Resources Canada announced that it had licensed ENERGY STAR for use in that country, primarily for new homes and for appliances and other products.† Under this arrangement, Canada's existing EnerGuide program of product ratings will also remain in place.

---

*U.S. Environmental Protection Agency, Energy Star press release, February 12, 2008, http://yosemite.epa.gov/opa/admpress.nsf/dc57b08b5acd42bc852573c90044a9c4/1e156a04a68baa30852573ed005bea4e!OpenDocument, accessed April 27, 2008.

†http://oee.nrcan.gc.ca/residential/energystar-portal.cfm, accessed May 17, 2009.

BOMA Canada BESt Program

BOMA BESt is a Canadian national environmental recognition and certification program for existing commercial buildings. There are four levels of certification in the program that incorporate BOMA Go Green best practices and the Go Green Plus assessment framework. According to BOMA Canada, "The Go Green program comprises two elements in which buildings and firms can choose to participate. While Go Green is a 'best practices' model, Go Green Plus adds a more in-depth benchmarking tool, for firms and buildings requiring or desiring this type of program."* The first level of certification recognizes that a building has met all of the Go Green best practices. Level 2 certified buildings have not only met that standard, but have also demonstrated a score of 70 to 79 percent on the Go Green Plus assessment. Level 3 certified buildings achieve a score in the 80 to 89 percent range, while Level 4 certified buildings score over 90 percent. BOMA BESt enables participants to develop action plans for energy, water, and waste reductions. These tools also help building owners and managers evaluate a portfolio of buildings and identify the strengths and weaknesses of each building. BOMA BESt is a valuable marketing and tenant relations tool that fosters increased environmental consciousness in building design and operations.

PROJECT PROFILE: ICAO BUILDING, MONTREAL, QUEBEC, CANADA

The International Civil Aviation Organization (ICAO) building received a LEED-EB Gold certification in December 2007, with a total of 48 out of 85 possible points. Built in 1996 and comprising more than 500,000 square feet of space, the project employed green exterior and site maintenance standards along with water-efficient landscaping. There was a strong focus on upgrading building management practices, along with energy performance measurement. The project also adopted all of the LEED-recommended green cleaning practices and focused considerable attention on sustainable purchasing policies and actions. More than 50 percent of the regularly occupied spaces have LEED standard daylighting, while more than 40 percent have views to the outdoors.†

LEED for Existing Buildings: Operations and Maintenance (LEED-EBOM)

The LEED-EBOM rating system helps building owners and operators measure operations, improvements, and maintenance on a consistent scale, with the goal of maximizing operational efficiency while minimizing environmental impacts. LEED-EBOM

---

*BOMA Canada, http://www.bomafest.com/about.html, accessed May 26, 2009.

†LEED Scorecard for this project, http://www.usgbc.org/LEED/Project/CertifiedProjectList.aspx.

addresses whole-building cleaning and maintenance issues (including chemical use), recycling programs, exterior maintenance programs, and systems upgrades. It can be applied both to existing buildings seeking LEED certification for the first time and to new construction or major renovation projects previously certified under LEED for New Construction (LEED-NC) or LEED for Core and Shell (LEED-CS). LEED-EBOM 2009 is the latest version of the LEED for Existing Buildings rating system, which originally debuted in 2002 as LEED-EB version 1.0. LEED-EB version 2.0 was used from 2004 through 2007, and LEED-EBOM was introduced in 2008.

Owing to the building upgrade component of LEED-EBOM, sometimes there may be confusion whether this rating system or LEED-NC is most appropriate for a particular project, since LEED-NC also applies to major renovations. A general guideline can be used: if the alterations or renovations affect at least 50 percent of the building's floor area or require removal of more than 50 percent of the tenants, LEED-NC is the most appropriate rating system.* Otherwise, LEED-EBOM should be applied. (*To avoid having to write LEED-EBOM throughout the rest of this book, I will just use LEED from here on. Where another LEED rating system is specifically referenced, I will use that system's full acronym.*)

CASE STUDY: SASAKI ASSOCIATES, WATERTOWN, MASSACHUSETTS

The noted architecture, landscape architecture, and planning firm, Sasaki Associates turned its Chase Mills headquarters (Fig. 4.3), which dates to the 1850s, into the oldest LEED-EB Gold building in Massachusetts and one of the oldest in the nation. The process of certification took more than two years and involved a major renovation of the building as well as incorporation of numerous sustainability components. Sasaki's Watertown offices operate as part of a larger initiative within Sasaki called GreenLAB, which uses Chase Mills as a laboratory for sustainability.†

Some of the LEED-EB initiatives at Chase Mills included: reducing building electricity loads by 30 percent by installing energy-efficient lighting; reducing water use for landscape irrigation; testing new green building materials; diverting 67 percent of waste generation from the landfill through use of "dumpster dives;" and reducing single-occupant commuting 30 percent by offering bike shelters, bike purchase discounts, public transportation vouchers, and employee discounts for "rent by the hour" car use.

One way to remember the difference between LEED-NC and LEED-EBOM is that LEED-NC resembles a snapshot of the building's expected energy and environmental performance during construction and for the period immediately following occupancy. By contrast, LEED-EBOM is more like a movie of actual performance during continuing operations over the time of the "performance period." Where LEED-NC estimates future energy and water use, LEED-EBOM actually measures and reports them, compared with established standards such as ENERGY STAR and the current national plumbing codes.

---

*LEED Reference Guide for Green Building Operations and Maintenance, p. xxiii.

†Personal communication, Sasaki Associates, April 2009.

Figure 4.3 **Sasaki Associates' Chase Mills headquarters building cut energy use by 30 percent through the use of energy-efficient lighting retrofits.** *Robert Benson Photography, courtesy of Sasaki Associates.*

THE LEED-EBOM PERFORMANCE PERIOD

If LEED-EBOM is a "movie" of a building's ongoing environmental attributes, naturally the question arises, "Over what time period are you measuring the impacts and upgrades?" LEED has created the concept of the "performance period" to guide project teams in assessing their building. In general, the performance period must be at least three months, that is, one quarter. This is quite easy and quite logical. Most building owners and managers expect to produce and receive quarterly performance reporting. However, for energy use, a one-year performance period must be used. At the project team's option, the performance period may include data from up to 24 months prior to application for LEED-EBOM certification. To keep all data consistent, all performance periods for everything that is measured during the certification process must overlap by at least one week.*

LEED-EBOM AS A CONTINUOUS IMPROVEMENT TOOL

Michael Arny is a principal author of the LEED-EB rating system, having served as the chair of the LEED-EB Committee during its development from 2001 through

*LEED Reference Guide, op.cit., p. xix.

2005, supporting the system's creation and growth. He says LEED should be looked upon as a continuous improvement tool for building owners and managers.*

> For buildings that have energy challenges, one of the important things for people to understand is that they can proceed to do all of the LEED-EB actions that fit in their current budget. Because most LEED-EB actions are low-cost and no-cost, the results will deliver significant positive environmental benefits. Then when people have the budget, they can make any improvements they need to get their ENERGY STAR score, for example, up to the point where they meet the prerequisite. This is a perfectly fine way to use LEED. It is important to see the LEED-EB rating system as a scale that provides a framework for moving up the scale and getting into the certified levels when you can and then continue to move up over time. This is the way I think building owners can most productively use LEED as a continuous improvement tool.

Arny also believes that advancing with LEED-EB certification is an easy decision for a building owner to make, because it directly adds value in the form of reduced operating costs and participation in the growing brand recognition of LEED certification:

> Moving forward with LEED should be a pretty easy business decision to make because you can control how it affects your expenditures. You can implement all the low-cost and no-cost LEED-EB actions and then move as quickly or as slowly as you want on executing any actions that have significant costs. For leased buildings, there's going to be market demand for LEED-EB-certified space. If you have buildings that you lease out and want them to be perceived as top-quality, more and more, I think ongoing LEED-EB certification and recertification will be required to maintain that image and reputation.

LEED VERSION 3

Let's take a closer look now at what's in LEED-EBOM, or LEED 2009 for Existing Buildings: Operations and Maintenance, part of the family of rating systems encased in a larger package called LEED v3 (version 3). LEED v3 includes the LEED Online reporting and certification templates, as well as a new "certification model," using the Green Building Rating and Certification Institute (GBCI) to accredit LEED professionals. GBCI also administers the LEED certification system through 10 outside "certification bodies" that review documentation and issue project certifications.†

LEED-EBOM completed the separation between the rating systems for new construction and for existing buildings; previously, many of the LEED-EB credits and prerequisites had been derived from the new construction rating system and just didn't fit with a rating and certification systems for ongoing building operations and maintenance. In addition, LEED-EBOM reduced the number of prerequisites from 13 to 9 and eliminated some obsolete provisions that had hindered uptake of the LEED-EB system.

---

*Interview with Michael Arny, April 2009.

†www.gbci.org, accessed May 22, 2009.

More importantly, LEED-EBOM rationalized and reorganized many of the credits into logical categories such as purchasing policies and waste management policies, with much clearer accountability for results in the way buildings are typically managed.

CASE STUDY: WATT PLAZA, LOS ANGELES

In May 2009, Watt Plaza, a twin 23-story office tower complex encompassing 900,000 square feet in the heart of Los Angeles' Century City development, was awarded LEED-EBOM Gold certification, the first office building in Los Angeles to achieve this recognition and one of the first 12 buildings in the United States certified under LEED-EBOM (2008) (Fig. 4.4).* Watt Plaza installed low-flow faucets, low-flush toilets, and 88 water-free urinals—measures which save millions of gallons of water per year. The team has also implemented a waste program that diverts 70 percent of all building waste to a materials recovery facility in addition to installing electronic waste, battery, lamp, and ballast recycling programs. To reduce single-occupant auto commuting, the project created alternative transportation options, including bicycle racks, promotion of increased participation in ridesharing, and providing commuting information from various Web sites to building occupants. Energy efficiency measures encompassed the following: achievement of a high ENERGY STAR rating; installation of window films throughout building interior, saving about 1 ton of air-conditioning for each 100 square feet of window treatment; and installation of 2231 occupancy sensors.

Still, in most of its structure and operations, LEED-EBOM is the "familiar" LEED system that has been around since 1999, with five key categories of environmental concern and an additional category of "bonus" points awarded for innovative performance. The five key categories are:

- Sustainable Sites
- Water Efficiency
- Energy and Atmosphere
- Materials and Resources
- Indoor Environmental Quality

In the LEED 2009 system, these categories have been "weighted" according to the significance of their environmental impacts, using a life cycle assessment tool developed by the U.S. EPA and applied to the LEED system by the USGBC. According to the USGBC:†

> With revised credit weightings, LEED now awards more points for strategies that will have greater positive impacts on what matters most—energy efficiency and CO_2 reductions. Each credit was evaluated against a list of 13 environmental impact categories,

*Watt Plaza LEED press release, http://www.businesswire.com/portal/site/google/?ndmViewId=news_view&newsId=20090519006478&newsLang=en, accessed May 23, 2009.

†www.usgbc.org/DisplayPage.aspx?CMSPageID=1971, accessed May 17, 2009.

Figure 4.4 **Watt Plaza performed an extensive building upgrade, focusing on water conservation, improved waste recycling, and lowering energy bills through window treatments and occupancy sensors.**
Photo courtesy of Watt Companies.

including climate change, indoor environmental quality, resource depletion, and water intake, among many others. The impact categories were prioritized, and credits were assigned a value based on how they contributed to mitigating each impact. The result revealed each credit's portion of the big picture, giving the most value to credits that have the highest potential for making the biggest change.

Figure 4.5 shows the relative weightings of the key main categories of concern, and Table 4.2 shows the numerical points awarded for each category, including the

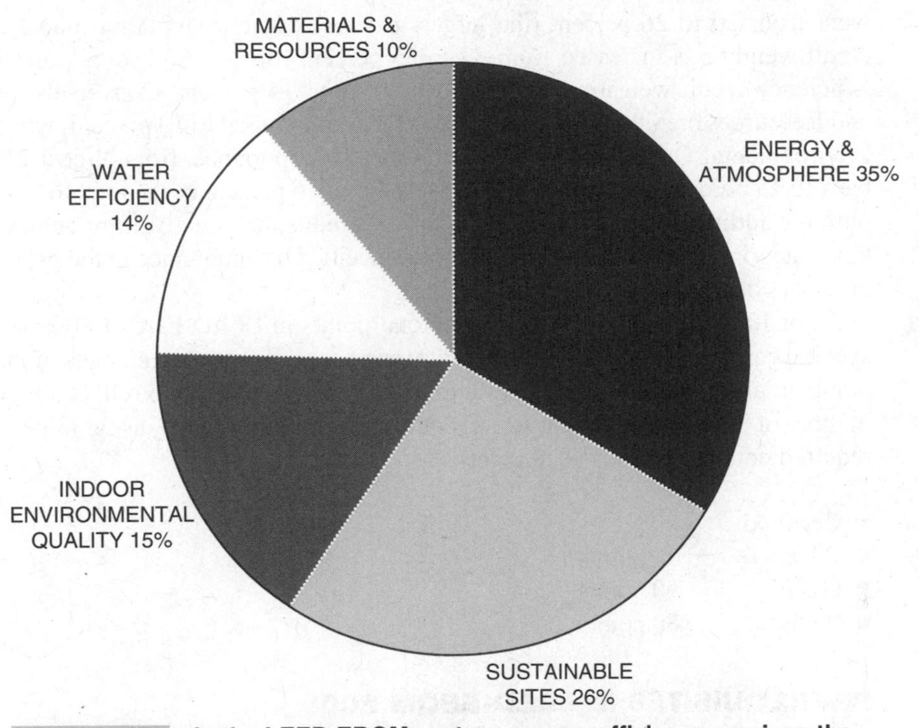

Figure 4.5 In the LEED-EBOM system, energy efficiency receives the greatest weighting, followed by sustainable site management.

"bonus" category. As we delve into each category in detail in later chapters, we'll expand the discussion of specific LEED credits within each of the broader categories of concern.

Several category weightings changed dramatically in the transition from LEED-EBOM (begun in 2008) to LEED-EBOM 2009. Sustainable Sites credit weighting

TABLE 4.2 LEED-EBOM 2009 POINTS IN EACH CATEGORY

CATEGORY	POINTS	PERCENTAGE OF TOTAL "CORE" POINTS
1. Sustainable Sites	26	26
2. Water Efficiency	14	14
3. Energy and Atmosphere	35	35
4. Materials and Resources	10	10
5. Indoor Environmental Quality	15	15
6. Innovation and Regional Priorities	10	n.a.
Total	**110**	**100**

went from 20 to 26 percent (the largest change). Energy and Atmosphere category credit weightings increased from 33 to 35 percent (of the 100 "core" points). Water Efficiency credit weightings increased from 11 to 14 percent. Conversely, Materials and Resources credits dropped from 15 percent of the total to 10 percent, while Indoor Environmental Quality credits also decreased in importance, from almost 21 percent back to 15 percent. Innovation credits went from 7.6 percent of the total to 9.0 percent, with the addition of 4 "regional importance" credits that are given for achieving specific category points in "regular" categories deemed of importance in the project's specific zip code area.

As of June 2009, it now takes more total points in LEED-EBOM 2009 to achieve specific certification levels. These points represent the same percentages of the "core" points in all of the other LEED systems, just adjusted upward to reflect a higher total number of such points. In that sense, nothing has really changed, as far as certification requirements go. Various achievement levels are

- Certified: 40 points
- Silver: 50 points
- Gold: 60 points
- Platinum: 80 points

PREREQUISITES IN LEED-EBOM 2009

One thing that tends to confound people new to the LEED rating and certification system is the presence of prerequisites, things you have to do in every project to achieve recognition. In LEED-EBOM 2009, there are nine prerequisites, shown in Table 4.3. In later chapters, we'll present these in more detail, so you'll have a good idea what you need to do to qualify a project for certification. For now, just know that you have to design your upgrades to meet these standards.

Rating Systems in Other Countries

The focus of this book is on greening existing buildings, with an emphasis on LEED-EB, the most widely used rating system in the world for existing buildings. However, the subject of certifying ongoing building performance is important worldwide, and some of the more established green building rating schemes have adopted a similar approach. In particular, consider the United Kingdom's "BREEAM" rating system, administered by the Building Research Establishment (BRE), which recently created a similar rating system.* BREEAM for new buildings antedates LEED by about five years and is still more widely used (on a per capita basis) than LEED. BRE now certifies "BREEAM In-Use" projects, to cover ongoing building operations.† The Green

*BREEAM is the acronym for the Building Research Establishment Environmental Assessment Method.

†www.bre.co.uk, accessed May 17, 2009.

TABLE 4.3 LEED-EBOM 2009 PREREQUISITES

PREREQUISITE	DESCRIPTION
1. Water efficiency	All fixtures to meet water efficiency requirements of the 2006 International Plumbing Code or the 2006 Uniform Plumbing Code, adjusted for age of building.
2. Energy and atmosphere #1	Adopt energy-efficiency best management practices for planning, documentation, and opportunity assessment.
3. Energy and atmosphere #2	Meet an ENERGY STAR score of at least 69; for projects not eligible for ENERGY STAR, improve efficiency at least 19% above average for similar types of buildings.
4. Energy and atmosphere #3	Zero use of chlorinated fluorocarbon (CFC)-based refrigerants in base building HVAC and refrigeration systems, unless economically infeasible.
5. Materials and resources #1	Adopt a sustainable purchasing policy and apply it to at least one major category.
6. Materials and resources #2	Adopt a solid waste management policy and apply it to at least one waste category.
7. Indoor environmental quality #1	Meet the requirements of ASHRAE Standard 62.1-2007 for minimum outside air ventilation.
8. Indoor environmental quality #2	Prohibit smoking in the building and at least 25 feet from building entrances, outdoor air intakes, and operable windows.
9. Indoor environmental quality #3	Adopt a green cleaning policy addressing seven specific requirements.

Building Council of Australia is also adapting its well-established "Green Star" rating system for the purpose of rating and certifying existing building performance.* Both of these systems are described in detail in Appendix IV. One can expect that other national rating systems for new construction will evolve certification programs for existing buildings; in the next few years, these will likely include the German system (Das Deutsche Gütesiegel Nachhaltiges Bauen), the French HQE (Haute Qualité Environnmentale), the Japanese CASBEE (Comprehensive Assessment System for Built Environment Efficiency), and other systems in use in such countries as China, UAE, and India.†

*www.gbca.org.au, accessed May 17, 2009.

†Sec Jerry Yudelson, 2009, *Green Building Trends: Europe*, for a more detailed description of European green building rating and certification systems.

BREEAM IN THE UNITED KINGDOM

BREEAM is certainly the oldest established green building rating system in the world. In 2008, BRE Global introduced a new scheme called "BREEAM In-Use" to complement its rating system for new construction. Thomas Saunders of BRE Global says:*

> We've designed BREEAM In-Use so it should be possible to use it on all nondomestic buildings. We expect the biggest users at first are going to be offices and retail. There's a lot of interest from public buildings and we've also had interest from universities. We've been talking to large-scale landlords and investors as well as owner-occupied and government buildings as well.
>
> The business case is that it allows you to track performance and improvements so that you can monitor the impact of an investment, to prove the performance of a building. If you want to seek funding from a sustainably invested pension fund, at the moment there's no other way of measuring the sustainable performance of investments. BREEAM In-Use will enable people to measure the performance before and after the investment so they can actually demonstrate to their pension scheme holders how much of an improvement they've made—how much of a reduction in environmental impact they've made.

BREEAM is well regarded in Europe and in fact has been adopted as the European standard by the International Council of Shopping Centers (ICSC) for all new construction. It's likely as well that the ICSC will adopt BREEAM In-Use as the sustainable building rating system for existing retail buildings.†

GREEN STAR IN AUSTRALIA

In 2007, the Green Building Council of Australia (GBCA) began an "extended pilot" assessment of a new rating tool, Green Star—Office Existing Building. This rating system was developed to assist building owners assess the environmental merits of their existing or future assets. GBCA deems the tool suitable for any of the following purposes: real-estate portfolio profiling, evaluation of assets for potential acquisitions, corporate environmental reporting, due diligence investigations, and identification of building upgrade priorities. The Green Star—Office Existing Building extended pilot evaluation continued through April 2009, as many project teams found it a little too complicated and documentation heavy.‡ The Office Existing Building program encourages and recognizes partial upgrades, up to 50 percent of a building's value, after which point it must use one of Green Star rating systems for new buildings.§ Even so, many projects upgrading older buildings have received Australian Excellence (5 Green Stars) and World Leadership (6 Green Stars) ratings.

---

*Interview with Thomas Saunders, April 2009.

†Personal communication, Ermine Amies, Director of ICSC Europe, London, May 2008.

‡Personal communication, Robin Mellon, GBCA, April 2009.

§www.gbca.org.au, accessed May 23, 2009.

GLOBAL CONVERGENCE OF RATING SCHEMES

The three leading green building rating schemes from the United Kingdom, United States, and Australia agreed in March 2009 to work together to develop common standards for measuring the environmental impact of new and existing buildings.

Under the terms of the agreement, the three rating tools (BREEAM, LEED, and Green Star) will set up a working group dedicated to mapping and developing "common metrics to measure emissions of CO_2 equivalents from new homes and buildings." The aim of the alliance is to ensure that multinational firms operating in different territories can easily compare the performance of their green buildings.

This alliance will ensure the carbon and energy data each tool uses is measured in the same way, making it easier to work out how buildings around the world are performing.*

BUILDING ENERGY LABELING IN THE EUROPEAN UNION

In 2002, the European Union (EU) adopted the Energy Performance in Buildings Directive (EPBD). The Directive, now moving toward full implementation, is an important step toward labeling all commercial and institutional buildings, something that will most likely appear in the United States within the next few years. (In fact, it is already a requirement in California, beginning in mid-2010.) It is based on a similar principle to ENERGY STAR: compare building energy use to a national average and give it a grade or a score. The difference is that while ENERGY STAR is still voluntary, by 2010 all EU states are supposed to have implemented the EPBD. One of the key driving forces for European energy-efficient design is the Directive.† Each of the 27 member states of the European Union is responsible for individual implementation of the EPBD through national laws. The main focus of European sustainable building design at this time is on reducing energy use directly and carbon emissions indirectly. Appendix II provides more details on the EU Directive.

Summary

What is a green building? In this chapter, we looked first at the major elements of green building, starting with the U.S. EPA's ENERGY STAR label, which assesses 15 different building types according to their annual energy use, compared with their peers. Then we examined the LEED system, with a focus on LEED-EB, the major established rating system for existing buildings. In this chapter, we provided an overview of the LEED-EBOM 2009 rating system, including the requirements (prerequisites) for all projects. In later chapters, we will delve into specific measures that can be implemented to achieve various certification levels.

---

*www.businessgreen.com/business-green/news/2237624/rival-green-building-codes-sign, accessed May 26, 2009.

†Jerry Yudelson, *Green Building Trends: Europe*, 2009, Washington, D.C.: Island Press, pp. 26–27.

5

THE BUSINESS CASE FOR GREENING EXISTING BUILDINGS

Within a very short period of time, the business case for new green buildings has become the "New Normal," considered "business as usual" by many major national property development, management, and ownership interests. Making the business case is vital; people in business have direct responsibility for managing economic assets for a specified return. "Saving the world" is generally not thought of as a priority by most businesspeople, though many organizations do manage to the "triple bottom line" presented in Chap. 1. In the new world of sustainable business, of course, good business and a healthy environment are not seen as mutually exclusive by any means. Still, where the rubber meets the road in business is profitability over both the short term and long term. Only then is a business truly "sustainable." Fast approaching is a time when all property owners and facility managers will consider the business case for greening existing buildings as proven. That time is not quite here, but the experience and viewpoints of some of the leading practitioners of energy-efficient and green building operations are helping to usher in the New Normal.

Major Commercial Developers

Jerry Lea, senior vice president of the international real estate firm Hines, a strong proponent and developer of both ENERGY STAR and LEED buildings, says, "I think sustainable is here to stay. The definition of 'Class A' buildings very soon will include sustainable design and probably LEED certification."* Adam Rose is the property manager at 717 Texas, a Hines office development in Houston. He agrees, saying that a commitment to sustainable operations demonstrates a firm's philosophy in practice.†

*Interview with Jerry Lea, May 2007.
†Interview with Adam Rose, April 2009.

We believe as a firm that a sustainable approach to real estate makes good business sense and is the right thing to do. Using an industry-standard rating system like LEED validated what Hines has been doing for years—we believe that a Class A building in the future is going to be defined by efficiency and sustainability: You're not going to have a Class A building that is not lean and green.

As far as [an increase in] the value of the building, it's hard to say because we haven't refinanced or sold this property since we developed and certified it. We feel like there's a definite value to the building being LEED-EBOM certified, and especially to being one of the very first to have that designation. We think it tells a strong story that the building is well-designed, well-managed and is a great place to work.

Jack Beuttell is sustainability manager at Hines. He thinks that a LEED-EBOM certification is definitely an asset for a commercial building such as 717 Texas, delivering a strong competitive advantage.*

The certification would give us a competitive advantage if at some point in the future we decided to market the property. It's a signal to tenants that the building is a great place to work. Higher occupancies and rents translate into higher market values. And those are things that catch an investor's eye.

Andrew McAllan is senior vice president of real estate management, at Oxford Properties Group in Toronto, the firm responsible for 255 King Street West in Toronto, the first multitenant LEED-EB-certified building in Canada.† He says his firm sees the business case as compelling for its future.

First of all we saw this holistically, that this is rapidly going to become an integral component of being a first-class office building. As one of the largest owner-managers of first-class office buildings and having a number of what are considered flagship properties across the country [Canada], it was just a must-have situation. We did some back-of-the-envelope calculations, but it's a challenge to empirically prove the payback calculation in advance. Rather, it's more about the sensitivity analysis. In this case *we concluded that if it assisted us in leasing the vacant space—it was an average 6 months earlier than we would have otherwise leased it—then just on that alone, it paid for itself.*

Oxford Properties also knew it had to compete with new buildings that would carry the LEED-certified label. This realization is rapidly gaining adherents in corporate real estate circles.

It's no coincidence that new buildings in downtown Toronto all have LEED certification. I'm hesitant to just categorize it as marketing because it's a lot more than marketing; it's a business decision and it's also part of our corporate social responsibility. It's part of Oxford's industry-leading Sustainable Intelligence program, where we are the first office and commercial real estate company in Canada to calculate in detail our greenhouse gas

---

* Interview with Jack Beuttell, April 2009.

† Interview with Andrew McAllan, April 2009.

> emissions, publish them and then set a target of reducing them by 20 percent by 2012. That's indicative of the corporate social responsibility that we have around these issues and therefore, being LEED-EB certified was a very important element. When we achieve LEED-EB certification, we are getting the seal of approval from an independent group, saying this property employs appropriate practices, systems, and equipment, etc.

Here Andrew McAllan expresses also the benefit of endorsements by recognized third parties such as the U.S. Green Building Council (LEED-EB) or the U.S. EPA (ENERGY STAR) in validating a building owner's claims about sustainable practices.

CASE STUDY: OXFORD'S METROCENTRE, 255 KING STREET WEST, TORONTO, ONTARIO

Oxford's MetroCentre earned the distinction of being the first multitenant LEED-EB certification in Canada in May 2008 with a Silver rating, achieving all 43 credits that it pursued in the certification process (Fig. 5.1). This property undertook the following innovative measures to secure the LEED-EB certification:*

- Converted its conventional air-conditioning system to an Enwave Deep Lake Water Cooling system, the world's largest lake-source cooling system
- Implemented enhanced operational policies that address site and grounds, pest management, green cleaning, waste recycling, maintaining air quality during construction, and sustainable purchasing
- Provided tenants with a suggested purchasing policy to encourage sustainable procurement practices
- Supplemented the available municipal transportation options with enhanced options, including bicycle storage and changing rooms, an in-building Zipcar (rent by the hour) program, and preferred/discounted parking for alternative fuel vehicles that run on some other fuel than gasoline.
- Extensive educational programs for employees, vendors, tenants, and management on operating, servicing, and using a "green" building, including group training, in-building displays, informational kiosks, and messaging in elevators.

As a result of taking these steps, there was a measured 13 percent energy reduction experienced (from 2006 to 2007), resulting in a 234-ton reduction in CO_2 emissions and an impressive ENERGY STAR rating of 95; interior lighting was programmed to operate only during regular business hours and the building reduced lighting loads by 20 percent. In addition, the building reduced waste hauled to landfills by 17 percent, with an overall 76 percent waste diversion rate for the entire building. The building's waste disposal policies mandate environmental disposal of toxic elements such as computers and batteries. The building also uses low environmental impact cleaning products, practices, and pest management. By refining the building's heating, ventilating, and air-conditioning (HVAC) system, tenants now receive 30 percent more outdoor air

*http://dcnonl.com/article/id31334, accessed July 24, 2009; personal communications, HOK architects, Toronto, May 2009.

Figure 5.1 **The office tower at 255 King Street West in Canada's business capital, Toronto, the first multitenant LEED-EB-certified project in Canada, is a great example of what can be done by private owners to upgrade the value of commercial properties.** *Photo by Shai Gill, courtesy of Oxford Properties Group.*

ventilation than the minimum required by LEED standards, improving indoor air quality for occupants. The certification process verified that existing plumbing fixtures conform to LEED water consumption requirements.

Benefits to Government Agency Owners

It's not only commercial building owners, developers, and managers who see benefits in LEED-EB certification, but also large property owners and managers in federal, state, and local governments. Johnathan Sitzlar is a property manager with the U.S.

General Services Administration. He says investing in LEED certification has many benefits for public ownership, citing the John J. Duncan Federal Building in Knoxville, Tennessee, which achieved LEED-EB certification in 2007.*

> It was for the betterment for the taxpayer, the betterment for the tenants in the building and the betterment of the building overall as far as extending its usefulness and peoples' desire to be in it. We maintained 100 percent occupancy and our customer satisfaction went up in the process. We went from 91 percent customer satisfaction to 93 percent. We're creating a better environment here in Knoxville and the taxpayer is reaping the benefits of the savings. We viewed it as a win-win for everyone and that's why we did it.
>
> We receive our capital expenditures from one pot of money regionally. In requesting funds for capital expenditure needs of the facility, we presented return on investment, cost savings, water savings and other analyses to show that we're maximizing our use of funds. That's our standard approach. Conscious effort to use this funding and good building management principles allowed for LEED-EB attainment. The capital expenditures were not developed specifically for attainment of LEED-EB but instead presented themselves based on forecasted and planned replacements of old equipment/systems. The GSA team viewed these projects as opportunities to maximize our value and improve energy consumption and sustainability.
>
> The LEED-EB certification has had a dramatic impact. GSA has received a lot of positive publicity associated with it. We had the opportunity to challenge other federal, state, local and private sector building managers and create new standards. It also challenged the private sector. We competed against both government and private sector facilities for the BOMA TOBY International Earth Award, which the building received in 2008. We have aided and been a stepping-stone in sustainability, environmental savings and energy conservation. We've had the opportunity to speak to Congress on behalf of this facility and with numerous industry colleagues throughout the country.

CASE STUDY: JOHN J. DUNCAN FEDERAL BUILDING, KNOXVILLE, TENNESSEE

In achieving LEED certification, the Duncan Federal building provided bicycle storage and shower/changing rooms for occupants, to encourage use of alternative transportation means (Fig. 5.2). Also to encourage alternative transportation, the building now provides preferred parking for alternative fuel vehicles. The building roof complies with both LEED and ENERGY STAR emissivity and reflectivity requirements, to reduce solar heat gain. Nearly 50 percent of the building hardscape (nonroof impervious paving) area is shaded, to reduce local heat build-up. Water use was reduced 40 percent, by installing waterfree urinals and low-flow aerators on lavatories, compared with the LEED baseline for the building. The project achieved an ENERGY STAR score of 94; in addition the project put in place an energy-efficient building systems maintenance and monitoring program, along with enhanced metering of

*Interview with Johnathan Sitzlar, April 2009. Also see, http://www.gsa.gov/Portal/gsa/ep/contentView.do?contentType=GSA_BASIC&contentId=22939, accessed May 24, 2009.

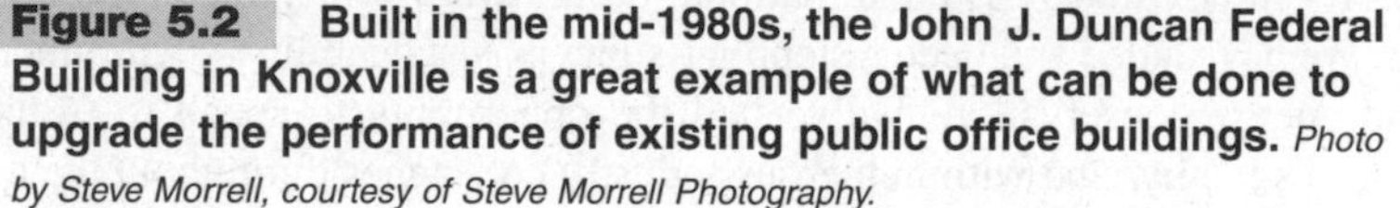

Figure 5.2 **Built in the mid-1980s, the John J. Duncan Federal Building in Knoxville is a great example of what can be done to upgrade the performance of existing public office buildings.** *Photo by Steve Morrell, courtesy of Steve Morrell Photography.*

energy use. Throughout the building, sustainable cleaning products and materials are used and green cleaning practices were adopted, including providing entryway mats and isolating the emissions from janitorial closets. The project also adopted a low environmental impact cleaning and pest management policy.

The Duncan Federal Building went through a comprehensive retrocommissioning of operating systems and procured a new Building Control System (BCS). The building operations staff uses the system to maximize efficiency of the HVAC system, establish cost-effective machine run times, use outside air to meet the demands of the building, immediately pinpoint problems in all systems, and produce numbers for trend data analysis that helps improve operating decisions.

Systems that were retro-commissioned and connected to the BCS included lighting on all floors, all variable-frequency drives, chillers, boilers, cooling towers, garage exhausts, restroom exhausts, air handling units, the building economizer, and all package and split heating and air-conditioning units, as well as the domestic hot water supply loop.

Pre-retrofit lighting consisted of magnetic ballasts and F40 bulbs. Using a favorable life cycle cost analysis, GSA obtained funding for upgrading the building's

1536 lighting fixtures to electronic ballasts and T8 lamps, and for replacing incandescent lamps with compact fluorescents. Also, the team installed motion detectors to control the lighting in restrooms, conference rooms, and other nonregularly occupied spaces.

To further reduce energy and water consumption, the project had to identify the most environmentally sound means of upgrading the building's antiquated existing cooling tower, already scheduled for replacement. The existing cooling tower used two 15-horsepower motors and lost excessive amounts of water due to overspray. The new tower incorporates premium efficiency fill, and uses two premium efficiency 7.5-horsepower motors. Together, they have significantly reduced total operating time, and thus the electrical load on the building.

Through office waste recycling programs, construction material recycling, and educational programs for the tenants, the Knoxville team reduced waste destined for landfills by 40 percent. In 2005, for example, the team diverted 13.75 tons of white paper and over 300 pounds of aluminum from the landfills.*

Don Horn is director of the Sustainability Program in the Office of Federal High-Performance Green Buildings of GSA's Public Buildings Service. He has a simple rationale for the business case for LEED: it helps you to track and report what you're doing, use the results to improve operations, and show the benefits to your superiors, clients, and the congress.

> The LEED rating system provides documentation that says you are actually doing what you are saying you're doing. Many of the policy-type things with LEED-EB, such as green cleaning, integrative pest management, green purchasing, represent policies we already have in place within the GSA. But to have some actual measure that it's being implemented within an individual building is the value of LEED certification that we don't have otherwise—*the tracking and reporting mechanism.* Of course, it's similar with ENERGY STAR—you have the assurance that you're operating your building in an energy-efficient manner to save on operating costs. In general, we have found that most of our federal buildings are operated far better than average commercial office buildings. We have a high percentage of buildings that are ENERGY STAR-labeled. Even for those that are not, if we actually go through the exercise to check the ENERGY STAR score, they usually score fairly high. The building that my office is in was built in 1932 and has an ENERGY STAR score of 98. The benefits of rating systems are the documentation of achievement and the operational cost savings.

For a long-term owner-operator such as GSA, it makes all the sense in the world to create healthy, productive, and energy-efficient buildings for their tenants, federal agencies. After all, the majority of costs in any office building derive from occupants, often dwarfing energy costs by a ratio of 100 to 1 (an average of perhaps $300 per square foot per year for salaries and benefits, vs. $3.00 for energy), so it makes sense to maximize occupant health and productivity.

---

*ww.gsa.gov/gsa/cm_attachments/GSA_DOCUMENT/Sustainability_Matters_508_R2-mQC1_0Z5RDZ-i34K-pR.pdf, accessed July 6, 2009.

The Multifold Dimensions of the Business Case

The business case for greening existing buildings is based on a framework of benefits: economic, financial, productivity, risk management, public relations and marketing, and project funding, some of which were outlined in Chap. 3.* In Table 5.1 you'll see a list useful for understanding the wide-ranging benefits of green buildings, which the

TABLE 5.1 BUSINESS CASE BENEFITS OF GREEN BUILDINGS

1. Utility cost savings for energy and water, typically 25 to 40 percent, along with reduced "carbon footprint" from lower energy use
2. Maintenance cost reductions from commissioning and other measures to assure proper HVAC systems performance
3. Increased value from higher net operating income (NOI) and better public relations, owing to higher rents and greater occupancy in certified buildings
4. Tax benefits for specific green building investments such as those specified in several pieces of state and federal legislation since 2005*
5. More competitive real-estate holdings for private sector owners, over the long run, especially in comparison with LEED-certified new buildings
6. Productivity improvements for tenants
7. Health benefits for tenants, including reduced absenteeism
8. Risk mitigation, including lower tenant exposure to irritating or toxic chemicals in building materials from renovations or remodels, or cleaning practices
9. Marketing benefits, especially for developers and building owners
10. Public relations benefits, especially for developers, building owners, and managers
11. Recruitment and retention of key employees and higher morale, both for tenants and for building owners and managers
12. Increased availability of both debt and equity funding for building sales and for upgrading building performance to higher green standards
13. Demonstration of commitment to sustainability and environmental stewardship; shared values with key stakeholders

*See the discussion of these pieces of federal legislation in Chap. 7.
Source: Yudelson Associates.

---

*U.S. Green Building Council, Making the Business Case for High-Performance Green Buildings (Washington, D.C.: U.S. Green Building Council, 2002), available at https://www.usgbc.org/Docs/Member_Resource_Docs/makingthebusinesscase.pdf, accessed May 6, 2007. See also *Environmental Building News*, 14, no. 4 (April 2005), available at www.buildinggreen.com, accessed May 6, 2007.

following sections examine in detail. Some benefits accrue directly to the building occupants, some to the property owner or manager, and some to the building's future financial and economic performance.

Benefits Directly to the Building Owner

Let's examine first the easiest justification for greening an existing building: direct economic benefits to the building owner through reduced operating costs, higher rents, greater occupancy, and higher resale value.

ECONOMIC AND FINANCIAL BENEFITS

Increased economic benefits are the prime drivers of change for most innovations; for energy-efficient and green buildings, these benefits take a variety of forms, and their full consideration is vital for promoting any sustainability initiatives.

Reduced operating costs With electricity prices rising steadily in many metropolitan areas, energy-efficient buildings make good business sense. In "triple-net" leases in which the tenant pays all operating costs, landlords still want to offer tenants the most economical space for their money. For a small investment in capital cost, green buildings can save on energy operating costs for years to come. In an 80,000-square-foot building, using $3.00 per square foot per year of energy, an owner's savings of 25 to 40 percent translates into $60,000 to $96,000 per year of reduced operating costs, year after year.

In a public setting, it's more difficult to get the funds to make needed upgrades and improvements. Barry Giles, formerly a facility manager at the California State University's Moss Landing Marine Lab, certified LEED-EB in 2004,* says:

> Return on investment was the first thing because we're part of the California State University system. In that system, it really has to show some financial benefits within three to five years. Normally, we do a matrix on costs versus return on investment, showing the return on investment within a specific time span. [A few years ago], we were one of the pilot projects for LEED-EB. We went at it absolutely full-tilt. Even though we didn't have to spend a dollar on the structure of the building, we still got a Gold rating for LEED-EB.†

Reduced maintenance costs More than 120 studies reviewed by Lawrence Berkeley National Laboratory have documented that an energy-saving new building, if properly commissioned, will show additional savings of 10 to 15 percent in energy costs, compared with one that has not been. Retrocommissioning will yield 5 to

*http://www.usgbc.org/LEED/Project/CertifiedProjectList.aspx, accessed May 26, 2009.

†Interview with Barry Giles, Moss Landing Marine Laboratory, Moss Landing, California, May 2007.

10 percent annual energy savings, according to a 2008 survey.* These buildings also tend to be much easier to operate and maintain, something discussed at greater length in Chap. 7. These benefits include extended equipment life, greater thermal comfort, and improved indoor air quality.† By conducting comprehensive functional testing of all energy-using systems in normal operations, it is often possible to have a smoother-running building because potential problems are identified on a regular basis. Retro-commissioning of commercial buildings is a formal way to examine potential energy savings improvements and to upgrade a building's ENERGY STAR rating.

Tax benefits Many states offer tax benefits for green buildings, including tax credits, tax deductions, property tax abatement, and sometimes sales tax relief. For example, New York's tax credit allows builders who meet energy goals and use environmentally preferable materials to claim up to $3.75 per square foot for interior work and $7.50 per square foot for exterior work against their state tax bill. In rehabilitated buildings, energy use cannot exceed 75 percent of the amount allowed under the New York State's energy code.‡

The 2005 Federal Energy Policy Act (EPACT) and subsequent amending legislation offer two major tax incentives for greening existing buildings: first, a tax credit of 30 percent on installed cost of both solar thermal (water heating) and solar electric (photovoltaic) systems, good through the end of 2016; second, a tax deduction of up to $1.80 per square foot for projects that reduce energy use for lighting, HVAC, building envelope measures, and water heating systems by 50 percent compared with a 2001 baseline standard, good through the end of 2013.§ This is a great incentive for energy conservation: For example, on a 300,000-square-foot commercial office building, a tax deduction of $540,000 is available, netting a potential tax savings of $135,000 at a 25 percent marginal federal tax rate.

Other incentives Depending on where you're located, there may be a number of other financial and project incentives for green building investments. Here are a few that you might look for:

1 State tax credits and sales tax exemptions in various states on material purchases¶
2 Property tax exemptions
3 Utility cash rebates, grants, and subsidies (typically based on energy savings and/or use of renewable energy systems)

---

*http://www.fmlink.com/News/Articles/news.cgi?display=article&id=25121, accessed May 24, 2009.

†Lawrence Berkeley National Laboratory, The Cost-Effectiveness of Commercial-Buildings Commissioning, 2004 [online], http://eetd.lbl.gov/emills/PUBS/Cx-Costs-Benefits.html. This research reviewed 224 studies of the benefits of building commissioning and concluded that based on energy savings alone, such investments have a payback within five years.

‡http://www.dec.ny.gov/energy/1540.html, accessed July 6, 2009.

§U.S. Department of Energy [online], www.energy.gov/taxbreaks.htm, accessed March 6, 2007.

¶For a list of state tax incentives for renewable energy, see the Directory of State Incentives for Renewable Energy, www.dsireusa.org; other local government incentive programs can be found at https://www.usgbc.org/ShowFile.aspx?DocumentID=691.

4 Permit assistance, including faster permitting or priority processing for major renovations (varies by jurisdiction)
5 Increased financing from socially responsible investors, such as pension funds and green building-focused real estate investment trusts (REITs) and private investment groups, sometimes facilitated through organizations such as the Clinton Climate Initiative

There is a reliable and fairly complete source of current information on all types of incentives in the Database of State Incentives for Renewable Energy and Energy Efficiency, available from the North Carolina Solar Energy Center.*

PRODUCTIVITY GAINS

In the service economy, productivity gains for healthier indoor spaces are worth anywhere from one to five percent of employee costs, or about $3.00 to $30.00 per square foot of leasable or usable space. This estimate is based on average employee costs of $300 to $600 per square foot per year (based on $60,000 average annual salary and benefits and 100 to 200 square feet per person).† With energy costs typically less than $3.00 per square foot per year, it appears that productivity gains from green buildings could easily equal or exceed the entire energy cost of operating a building. For owner-occupied buildings, such as corporate, education, industrial, or institutional facilities, all of the gain in productivity accrues to them.

Here's an example, shown graphically in Fig. 5.3. Median productivity gains from high-performance lighting of 3.2 percent in 11 studies were reported by Carnegie-Mellon University in Pittsburgh, or about $1.00 to $2.00 per square foot per year, an amount equal to the cost of energy.‡ This is in addition to a reported average savings of 18 percent on total energy bills from proper lighting. For corporate and institutional owners and occupiers of buildings, that is too much savings to ignore. For landlords, it's important to realize that tenants are beginning to realize productivity gains associated with green buildings. For this reason, buildings using LEED-EB to improve daylighting integration and secure better indoor air quality (often through changes suggested by retrocommissioning audits) can provide important benefits for a company or organization.

Additionally, a recent study by the Center for the Built Environment at the University of California, Berkeley, that reviewed more than 33,000 surveys of occupant satisfaction in more than 200 buildings, including 16 green-certified buildings, showed a statistically significant gain in certified green buildings, compared with

*www.dsireusa.org, accessed May 23, 2009.

†Eleven case studies strongly suggest that innovative daylighting systems can pay for themselves in less than one year due to energy and productivity benefits. Vivian Loftness et al., Building Investment Decision Support (BIDS) (Pittsburgh: Center for Building Performance and Diagnostics, Carnegie Mellon University, n.d.), available at http://cbpd.arc.cmu.edu/ebids, accessed March 6, 2007.

‡Carnegie Mellon University, http://cbpd.arc.cmu.edu/ebids/images/group/cases/lighting.pdf, accessed March 6, 2007.

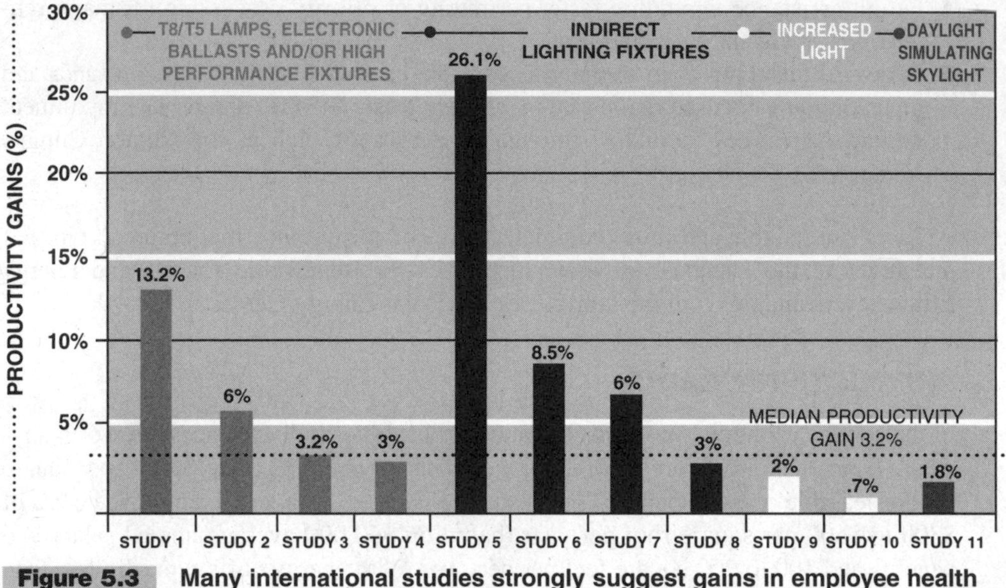

Figure 5.3 **Many international studies strongly suggest gains in employee health and productivity from improving the quality of interior lighting and integrating natural daylighting with artificial lighting.** *Center for Building Performance and Diagnostics, Carnegie Mellon University. eBIDS™: Energy Building Investment Decision Support Tool.*

those that were not certified, considering air quality and thermal comfort. In the words of the researchers: "Our results suggest that on average the strategies commonly employed in green buildings have been effective in improving occupant satisfaction with air quality and thermal comfort."* Building operators should then ask themselves: What could higher occupant satisfaction do to improve productivity, reduce health costs, lower employee turnover, and improve morale in any organization?

RISK MITIGATION

Risk in building operations has multiple dimensions: financial, market, legal, and reputational. Since it's often hard to increase net operating income from building operations in the short run, mitigating risk exposure also has positive economic benefits.

Sick building syndrome LEED-EB certification, along with retrocommissioning audits that lead to improvements in indoor air quality, may provide some measure of protection against future lawsuits through third-party verification of measures installed or upgraded that protect indoor air quality, beyond just meeting code-required minima. With a national focus on mold and its effect on building occupants, building owners should be paying more attention to improving indoor air quality.

*www.cbe.berkeley.edu/research/pdf_files/Abbaszadeh_HB2006.pdf, accessed May 27, 2007.

Here's a good example: a building refurbishment in Melbourne, Australia, of the 500 Collins Street building, documented the following benefits:

- 39 percent reduction in average sick leave days
- 44 percent reduction in the average cost of sick leave
- 7 percent increase in billings ratio, despite a 12 percent decline in the average monthly hours worked*

Meeting pro-forma projections Another risk management benefit of LEED-certified buildings in the private sector is the higher rents and greater occupancy of such buildings, compared to similar projects in the same city. In Chap. 3, we also presented growing evidence of considerably greater resale value for LEED and ENERGY STAR buildings. Green buildings tend to be easier to rent and sell, because educated tenants can increasingly understand and directly experience their benefits.

Reduced cost of commercial insurance Insurers also see green buildings as less risky. For example, in September 2006, Fireman's Fund, a major insurance company, announced it would give a five percent reduction in insurance premiums for green buildings. The insurer also announced its "Certified Green Building Replacement and Green Upgrade" coverage.[†] In 2010, the National Association of Insurance Commissioners will begin requiring insurance companies to disclose their potential liability from climate change.[‡] At some point in the near future, insurers are very likely going to be posing similar questions to building owners.

Refinancing a portfolio Denis Blackburne is chief financial officer of Melaver, Inc., a sustainably minded commercial real estate development company based in Savannah, Georgia. He says:

> One financial aspect of LEED certification is how it helped us when we refinanced some of our properties. We put our suits on and went to New York City and met with the top financial institutions and said, "We have a portfolio of six green properties for you." Basically they thought we were tree huggers and we were sent off to the "community grants" floor. We went back home, did our homework and presented our case to the financial community the second time around by saying, "We have six properties that have quality tenants, are in the right location, look great, are well-managed, have high occupancy and by the way they are high-performing, in the sense that they are energy efficient and environmentally friendly." All of a sudden all of the doors opened and we were able to refinance this portfolio and exceed our objective by far. When we looked at the future cash flow projections, we were given full credit for the future benefits that were included in the LEED certification.[§]

---

*http://www.resourcesmart.vic.gov.au/news_and_events/business_news_3945.html, accessed May 13, 2009.

†www.buildingonline.com/news/viewnews.pl?id=5514, accessed May 6, 2007.

‡http://www.naic.org/Releases/2009_docs/climate_change_risk_disclosure_adopted.htm, accessed May 24, 2009.

§Interview with Denis Blackburne, February 2008. See also, Scott Doksansky, "Existing Buildings: The Green-Headed Stepchild of the Sustainability Movement," pp. 167–198 in *The Green Building Bottom Line*.

HEALTH IMPROVEMENTS

Of course, a key element of productivity is healthy workers. By focusing on measures to improve indoor environmental quality such as increased ventilation, daylighting, views to the outdoors, and low-toxicity finishes and furniture, people in green buildings show an average reduction in symptoms of 41.5 percent on an annual basis, according to 17 academic studies reviewed by researchers at Carnegie Mellon University.*

Since most companies are effectively self-insured (i.e., your health insurance costs go up the more claims you have) and most government agencies and large companies are actually self-insured, it makes good economic sense to be concerned about the effect of building operations on people's health. Health claims are easy to measure statistically and to compare with local and regional averages.

MORE COMPETITIVE PRODUCT

Commercial building owners are realizing that LEED-certified buildings can be more competitive in certain markets. Green buildings with lower operating costs and better indoor environmental quality are inherently more attractive to a growing group of corporate, public, and individual tenants. "Green" will not soon replace known real-estate attributes such as price, location, and conventional amenities, but green features will increasingly enter into tenants' decisions about leasing space and into buyers' decisions about purchasing properties and homes.

Benefits to Both the Building Owners and Tenants

Many building owners, both public sector and private, are finding considerable benefits from green buildings, as we saw in the statements from Hines, Oxford Properties Group, and the federal GSA. These organizations saw benefits in the form of positive marketing and public relations with a variety of stakeholders.

PUBLIC RELATIONS AND MARKETING

Marketing is an essential component of all building operations, both public and private. Companies, public agencies, universities, and many nonprofit organizations seek to maximize their "brand equity" and increasingly rely on marketing and public relations activities to accomplish this goal.

*Center for Building Performance and Diagnostics, Carnegie Mellon University. BIDS: Energy Building Investment Decision Support Tool. http://cbpd.arc.cmu.edu/ebids.

Stakeholder relations and occupant satisfaction Tenants and employees want to see a demonstrated concern for their well-being and for that of the planet. Progressive building owners realize how to market these benefits to a discerning and skeptical client and stakeholder base, using the advantages of LEED and ENERGY STAR to indicate a positive response to a growing concern for the long-term health of the environment.

Mike Lyner is a principal of RSP Architects in Minneapolis. He believes that LEED-EB can help companies to demonstrate their sustainability commitments, even if they are already in a LEED-NC certified building:*

> Going forward with a new client, we certainly tell them about LEED-EB. It's reinforcing the fact that they're walking the talk, "You've got a LEED-NC rating, and you've done the best you can from Day One. Let's not lose momentum. Let's just keep it up. You do not want the building to start slipping backwards from what it's capable of doing. You would like to think with a brand new building that you would change the filters regularly, that you wouldn't put nasty paint on the walls that smells all day long. We've assisted you in creating a nice clean building to start with, don't mess it up." While they can claim a plaque going in, wouldn't it be nice to have a plaque alongside it saying that they kept it up according the standards conceived for the new building?
>
> Do green building certifications and attendant benefits actually help building owners reduce tenant turnover through greater satisfaction? Most leases in commercial real estate are net leases where the owner passes through all operating expenses to the tenant. Any savings produced go directly to the tenant, not the owner. Even though one could argue all day long that a happy tenant is a good thing and that should be motivation enough for owners' actions, it's tough to quantify what makes a tenant stay or leave. If operating costs [energy] are reduced, will a tenant renew or stay? It probably depends on many factors beyond just energy cost. But recall that one of the ongoing costs in operating buildings is paying for leasing commissions and tenant improvements [to meet the needs of new tenants], so if a tenant stays in a building, it's going to save the owner a reasonable amount of money. So, if upgrading energy performance can control these costs, it ought to work in the owner's favor over the long run.

Lyner's point is clear: a primary goal of every building owner is to reduce turnover; replacing tenants costs money, in lost income, fit-out expenses for a new tenant and marketing expenses. If a building has lower operating costs and better indoor environmental quality, over the long run, tenants are more likely to stay, and that's a good thing.

Environmental stewardship Being a good neighbor is appropriate not just for people or building users, but also for the larger community. It is a key component of "triple bottom line" management. Developers, large corporations, universities, schools, local government, and building owners have long recognized the marketing and public relations benefits of a demonstrated concern for the environment. Green

*Interview with Mike Lyner, May 2009.

buildings fit right in with this message. As a result, we expect to see major commitments by corporate real estate executives for greening their buildings and facilities. A good example is Adobe Systems, Inc., a major software maker based in San Jose, California. In 2006, Adobe announced that it had received LEED-EB Platinum awards for each of its three headquarters towers; not only did it reap great publicity, but the firm showed that the investments as a whole had returned a net present value almost 20 times the initial cost.*

Many larger public and private organizations have well-articulated sustainability mission statements and understand how their real estate choices can both reflect and advance those missions. Developer Jonathan F. P. Rose noted that "having a socially and environmentally motivated mission makes it easier for businesses in the real estate industry to recruit, and retain, top talent. Communities are more likely to support green projects than traditional projects, and it is easier for such projects to qualify for many government contracts, subsidies, grants, and tax credits. The real estate industry can prosper by making environmentally responsible decisions."†

Brand image Green buildings also reinforce a company's brand image. Consumer products companies such as Wal-Mart, Kohl's, Best Buy, and PNC Bank try to improve or maintain their brand images by being associated with green, solar, or energy-efficient buildings, and so many are moving in this direction.‡ Large corporations, including those issuing sustainability reports every year, are beginning to see benefits of building green that demonstrate to their employees, shareholders, and other stakeholders that they are "walking the talk."

Branding and Positioning a Commercial Property

A developer with a strong commitment to green projects and a growing portfolio of such projects might find an increased credibility in the tenant marketplace and an increased ability to win business from major corporate tenants with a strong commitment to sustainability. Jerry Lea of Hines says, "[green building] has definitely enhanced our reputation. We've always had good buildings in terms of energy conservation, indoor air quality and building systems. [Before LEED] nobody gave us credit for that or understood that. [LEED] has allowed people to understand that not only do our buildings look good, but they're good buildings to be in."§

---

*U.S. Green Building Council [online], www.usgbc.org/News/PressReleaseDetails.aspx?ID=2783, accessed May 6, 2007.

†"The Business Case for Green Building," *Urban Land*, June 2005, p. 71; www.uli.org.

‡For example, PNC Bank has committed to making all of its new branches LEED-certified, at least at the basic level.

§Interview with Jerry Lea, May 2007.

In other cases, green building certifications might present a golden opportunity to reposition older office and industrial properties as more upscale or trendy. Consider the benefit of certifying properties to LEED-EB or LEED for Commercial Interiors (LEED-CI) with each remodel or renovation. Such branding might help to make an older property more competitive with new LEED-certified buildings coming online in just about every major city. Alternatively, a developer could brand an entire office park as green using a large-scale photovoltaic (PV) solar energy system or tall wind turbine that would be architecturally iconic and highly visible. Establishing and improving the environmental performance of older properties could be an essential element in rebranding them and repositioning them to be more attractive to tenants looking for green office space.

RECRUITMENT AND RETENTION

One often-overlooked aspect of green buildings is the effect on people's interest in joining or staying with an organization. It costs an estimated 150 percent of annual salary to lose a good employee, and most organizations experience 10 to 20 percent turnover per year, some of it from people they didn't want to lose.* In some cases, people leave because of poor physical environments. In a workforce of 200 people, turnover at this level would mean 20 to 40 people leaving per year. What if a green building could reduce turnover by five percent, for example, one to two people out of the 20 to 40? Taken alone, the value of that would be $50,000 to possibly as much as $300,000, more than enough to justify the costs of certifying a building project. If a professional service firm, say a law firm, lost just one good attorney, typically billing $400,000 per year, with $250,000 gross profit to the firm, that could more than pay for the extra cost of a green building or green tenant improvement project that would keep that lawyer at the firm. Consider also the impact of a healthier work environment on employees' belief that their employer really cares about their well-being.

TOP-LINE REVENUE GAINS

Over the past 10 to 15 years, American business has focused profit-growth successfully on squeezing costs out of operations, through downsizing, rightsizing, outsourcing, logistics improvements, and so on. To grow profits now, companies need to grow top-line revenue. In the service economy, top-line revenue growth comes from recruiting and retaining top performers. The "80/20 rule" suggests that 80 percent of any firm's revenues will come from 20 percent of people or products, so companies have focused a good part of their marketing strategies necessarily on the "blocking and tackling" of making their companies attractive to key people.

Add to that the increasing scarcity of "Gen X'ers" (i.e., those born between 1965 and 1978) who increasingly occupy key project management and leadership roles in

---

*William G. Bliss, "The Cost of Employee Turnover," http://www.isquare.com/turnover.cfm, accessed July 6, 2009.

most organizations today (e.g., by 2014, people in the 35- to 44-year-old age group in the United States will be 7 percent fewer, a shortfall of 2.6 million people, than in 2005).* Demographics is destiny: one can see why green buildings might have such large potential importance—for many companies, they represent one of the few tangible and easily achievable measures for demonstrating a commitment to sustainability. This in turn helps them get and keep good people, especially those who drive top-line revenue growth (think of the importance, for example, of principals and senior associates in all manner of professional service firms—lawyers, accountants, architects, engineers—as well as in client contact and project leadership roles in finance, insurance, and real estate). This recruitment and retention issue is now emerging as a key rationale for most large organizations to consider greening all of their existing buildings. I believe the rationale is even more pronounced when considering recruiting and retaining the next generation of employees, the "Gen Y" group, which is passionate about environmental issues.†

Benefits to Future Financial Performance

Most private organizations have short-term planning horizons. But many large property owners have been in business for decades and plan to stay in business, so they must balance the short-term costs and benefits of greening their buildings with the longer-term positive outlook for sustainable buildings.

ENERGY COST REDUCTIONS

Energy costs represent 30 to 35 percent of a typical building's operating costs and certainly are among the more uncontrollable costs. Already, in major cities, commercial buildings are paying very high electricity rates for peak-period electricity. As electricity supply struggles to keep up with demand, one can expect utilities to try to reduce demand growth by increasing peak-period rates even more, trying to reduce peak-period use of summer air-conditioning. As we saw with the Empire State Building retrofit, demand reduction is a key strategy in extending the life of existing HVAC systems and in improving user comfort in older buildings. Anything you can do to reduce thermal loads in a building, such as installing reduced-wattage lamps, occupancy sensors, and daylight sensors, will also reduce peak-period electrical demand and the charges an owner has to pass through to the tenants.

*Jerry Yudelson, *The Green Building Revolution*, 2007, Washington, DC: Island Press, p. 41.

†http://www.canada.com/topics/technology/science/story.html?id=c24b5194-2277-4d75-b32d-0ff37d5eac59, accessed July 6, 2009.

Barriers to Realizing Green Building Benefits

In earlier chapters, we spoke of the drivers promoting LEED-EB and some of the inhibiting factors. Some of these arise whenever one is promoting the business case and deserve to be mentioned again in this context.

UNEQUAL DISTRIBUTION OF BENEFITS

One of the biggest issues in greening existing buildings is that the benefits are often unequally distributed between those who pay for the project and those who benefit. In a 2009 report, the World Business Council for Sustainable Development (WBCSD) drew the following conclusions about how the Intergovernmental Panel on Climate Change's (IPCC) goal (reducing world carbon emissions by 77 percent by 2050) might be met:*

1. Energy efficiency projects totaling $150 billion annually could reduce carbon footprints by 40 percent with five-year (or less) discounted payback periods, assuming energy prices around $60 per barrel of oil.
2. The study looked at six building markets that produce half of the world's GDP and generate two-thirds of global primary energy: Brazil, China, Europe, India, Japan, and the United States.
3. A further $150 billion annually (in energy efficiency projects), with paybacks between 5 and 10 years, could add another 12 percent in carbon footprint reductions, bringing the total to 52 percent.

In general, it's building owners who would have to put up the initial capital for these investments, but with current lease structures, it's mainly tenants who benefit. Yet these savings only make up one-third of the projections of what's needed to reduce building energy use to control carbon emissions. To go beyond those energy savings with shorter paybacks will require "integrated actions from across the building industry, from developers and building owners to governments and policy-makers."

Among other measures, tax incentives and subsidies would be needed to spur investments beyond a 10-year timeframe. Also needed would be updated building codes and inspections focused on energy efficiency, building renovations, and new projects to include integrated energy-efficient design and reduction of appropriate carbon costs.† Beyond these measures, the study calls for major cultural and behavioral

*"Energy Efficiency in Buildings—Transforming the Market," World Business Council for Sustainable Development, http://62.50.73.69/transformingthemarket.pdf, accessed May 1, 2009.

†The International Code Council (ICC) is a leading organization developing new building codes and initiatives focused on green buildings. See, for example, the new International Green Construction Code initiative between ICC, the American Institute of Architects and ASTM International (originally the American Society for Testing and Materials), www.bdcnetwork.com/article/CA6668304.html, accessed July 6, 2009. This initiative proposes to develop a model green construction code focused on new and existing commercial buildings.

shifts which can be the hardest thing to bring about, but are arguably the most effective in the long run.

Difficulties of implementing LEED-EB in K-12 schools In schools, the vast majority of green benefits accrue mostly to the students, but it's the school district that incurs the initial cost. (This is similar to the owner/tenant issue in commercial offices and retail buildings.) The school district has a primary interest in reducing operating costs, but the real long-term and significant benefits are in the health and productivity of students, teachers, and staff.* (One could argue of course that schools should be equally or more concerned about these health and productivity issues, but the author's experience is that school design is mostly about instructional needs, security, and budgets.) A further inhibiting factor is the fact that schools are strapped for operating funds even in the best of times, so trying to appropriate money for a LEED-EB certification can be quite difficult to justify. In this circumstance, the best solution might be to turn to energy service companies (ESCOs) and let a private party put up the capital for energy-related improvements, roll a LEED-EB certification into the program as an educational exercise, and let the future energy savings pay for the entire project.

Project profile: Flynn Elementary School, Eau Claire, Wisconsin As of May 2009, the first K-12 school in the United States had received LEED-EB certification, the John J. Flynn Elementary School (Fig. 5.4). The Eau Claire Area School District hired Ayres Associates architects to design a new, 54,000-square-foot, two-section elementary school, with flexibility to add a future third section. The school district was very interested in incorporating sustainable features into the project. Ayres Associates responded with an effective design that helped the district later receive a LEED-EB Silver certification. The original building design included high ceilings and extensive natural lighting for visual appeal and improved energy efficiency. By using daylight in instructional areas, the need for artificial lighting was minimized, and the size and expense of the air-conditioning equipment was reduced.†

Table 5.2 shows the distribution of green building benefits; when promoting green building upgrades, owners, and managers should always consider these distinctions in tailoring their case to key decision-makers. In the future, we expect public policy for promoting green buildings to take this unequal distribution of benefits into account by creating incentives to overcome these gaps in the marketplace. Some cities that have committed to reducing overall carbon emissions are already considering mandates for commercial building owners that would require them to get LEED-EB certification with any major renovation, an approach similar to what has been required on the West Coast in many cities, for example, for earthquake protection.

---

*"Greening America's Schools: Costs and Benefits," October 2006, Gregory Kats and Capital-e, downloadable from http://www.cap-e.com/publications/default.cfm, accessed May 23, 2009.

†Personal communication, Ayres Associates, April 2009. Information also at http://www.ayresassociates.com/news_events_articles_corporate.html, accessed May 23, 2009.

Figure 5.4 **Flynn Elementary achieved LEED-EB Silver in 2008 by coupling an energy-efficient base building with a strong focus on green cleaning programs and environmentally preferable purchasing policies.** *Mark Fay/Faystrom Photo, courtesy of Ayres Associates.*

TABLE 5.2 DISTRIBUTION OF LEED-EB OR ENERGY STAR BUILDING BENEFITS

BENEFIT TYPE/ OWNER TYPE	ENERGY SAVINGS	PRODUCTIVITY GAINS	HEALTH BENEFITS	MARKETING/ PR	RECRUITMENT
Private, owner-occupied	Yes	Yes	Yes	Yes	Yes
Private, speculative, or not owner-managed	No (for a typical "triple net" lease)	No	No	Yes	Maybe, but indirect
Retail	Yes	No	Maybe	Yes	No
University	Yes	Yes	Yes	Yes, very important	Somewhat important
Federal government	Yes	Yes	Yes	Not too important	Not too important
State government	Yes	Yes	Yes	Not too important	Not too important
Local government	Yes	Yes	Yes	Somewhat important	Somewhat important

Source: Yudelson Associates.

COST OF UPGRADES

Energy-efficiency and LEED-EB investments of $1.00 to $2.00 per square foot still represent significant capital outlays for many organizations. With low prevailing interest rates, it can be financially prudent to borrow the money for upgrades (if it is available). The capital available for financing these upgrades, while currently difficult for many companies to obtain, may become more accessible in 2009 and 2010 as economic conditions improve. For example, in 2007, before the "credit crunch" of 2007 and 2008, both Bank of America and Citibank announced major new green lending initiatives,* and many other large banks were expected to follow suit. In another case, to support former President Bill Clinton's "Clinton Climate Initiative," in May 2007, Citibank, Deutsche Bank, JP Morgan Chase, UBS, and ABN Amro each committed $1 billion to finance energy conservation upgrades for public buildings in 16 cities around the world.† Bank of America's initiative includes committing $1.4 billion to achieve LEED certification in all new construction of office facilities and banking centers and investing $100 million in energy conservation measures for use in all company facilities.

GREEN LEASES

In 2008, the Building Owners and Managers Association (BOMA) International released a model green lease. LEED-EBOM can be a powerful impetus toward rewriting standard leases to make them green leases. A green lease would address various LEED criteria, for example, controls on the materials and methods used in tenant improvements, requirements for minimum energy performance in remodels, and similar issues.‡

The Business Case for Greening the Retail Sector

The retail sector shares many of the same issues for LEED-EB as does the office sector. One can think of a retail mall as a horizontal (instead of vertical) multitenant space. One primary difference is that most of the building occupants during the course of a day are shoppers not store employees; another is that operating hours are about 40 percent greater than those of a normal office building.

In some cases, such as "big box" retail, the tenant has a ground lease and builds out their own space completely. Clearly, developers and retailers have common interests: to get as many consumers to a center to shop, as often as possible. They also have divergent interests: the developer functions primarily as a landlord, developing the property, maintaining it and collecting rents. At some point, the developer will likely sell the

---

*www.reuters.com/article/bankingfinancial-SP/idUSN0844820420070508, and http://bankofamerica.mediaroom.com/index.php?s=press_releases&item=7697, accessed May 27, 2007.

†http://www.london.gov.uk/view_press_release.jsp?releaseid=11975, accessed July 3, 2009.

‡http://shop.boma.org/showItem.aspx? product=GL2008&session=10438DEDC9CA4C48B16D687983ADC75D, accessed July 24, 2009.

property. The developer aims to collect enough rent to meet investor's goals for a return on investment and to keep the center as full as possible, for as long as possible.

The retailer's goal is to meet or exceed sales targets by satisfying consumer demand for products and services, while paying as little rent as possible. The retailer typically pays directly for utilities and indirectly through common-area-maintenance (CAM) charges, which it wants to keep as low as possible.

Because many consumers care about energy use and environmental impact, both developer and retailer often have common interests around green certifications such as LEED-EB. Just as office tenants can certify their individual spaces to LEED-CI standards, so can retailers. Additionally, freestanding retail buildings can be certified as new construction projects to LEED-NC standards. To date, other than the Stop & Shop grocery chain (Chap. 3, Case Study: Stop & Shop), most retailers and developers have not committed to the LEED-EB system, the best system for dealing with existing buildings. However, over the next three years, this situation is likely to undergo profound change, as the benefits of green certification become more apparent.

How to Implement LEED-EBOM

If you're a facility manager or building manager, the first step to reaping these benefits is to present the business case to your CFO or client and get the company to commit to LEED-EBOM certification across the existing building portfolio. As we have seen throughout this book, LEED is a comprehensive tracking, evaluation, and benchmarking system that will help you "green" your operations and engage your entire workforce in the effort. As of May 2009, more than 2500 projects using LEED-EB were in progress (with 213 certified), involving more than 930 million gross square feet of building area, an average of 366,000 square feet per project.* Commitments to using LEED-EB are growing with both private sector and public/nonprofit sector project registrations. In the private sector, in 2008 and 2009, CB Richard Ellis submitted more than 225 managed properties for LEED-EB certification.† In the public/nonprofit sector, in 2006 the University of California, Santa Barbara, agreed to evaluate 25 campus buildings against the LEED-EB standard over a five-year period.‡ In June 2008, the campus recreation center became LEED-EB Silver certified, the fourth building at the campus to achieve LEED-EB.§

PROJECT PROFILE: UNIVERSITY OF CALIFORNIA, SANTA BARBARA (UCSB), RECREATION CENTER

The Recreation Center (Fig. 5.5) reduced its measured environmental footprint in several ways: recycling was increased by 70 percent, trash sent to landfill was reduced by 55 percent and electricity, natural gas, and water consumption were reduced substantially,

*Calculated from USGBC data furnished to the author, May 2009.

†http://www.cbre.com/USA/Sustainability/Sustainability, accessed July 24, 2009.

‡University of California, Santa Barbara, December 17, 2006 press release, www.universityofcalifornia.edu/sustainability/documents/ucsbnews.pdf, accessed May 24, 2009.

§http://www.ia.ucsb.edu/pa/display.aspx?pkey=1790, accessed May 24, 2009.

Figure 5.5 **The first recreation center on any university campus to achieve LEED-EB recognition, the center increased recycling by 70 percent and reduced trash sent to landfill by 55 percent.** *Photo courtesy of University of California, Santa Barbara.*

all without reducing its operational capabilities. In addition, the Recreation Center will be installing a 133-kilowatt photovoltaic array on its Multi-Activity Center (MAC) roof in the summer of 2009. The solar array will be the largest on-site renewable energy generator on the UCSB campus, producing approximately 70 percent of the power required to run the 55,000-square-foot building.

The Recreation Center planned in 2009 to install solar thermal water heaters to provide preheated water for its 1.8-million-gallon pool, which, in tandem with the newly installed pool covers, should reduce natural gas usage by 80 percent and save $160,000 each year in fuel costs.

PROJECT PROFILE: CALIFORNIA EPA BUILDING, SACRAMENTO, CALIFORNIA

The 25-story, 950,000-square-foot California Environmental Protection Agency (Cal EPA) building was built in 2000 and is owned and operated by Thomas Properties Group. In 2003, Thomas Properties certified the Cal EPA building as the first LEED-EB Platinum-rated project.* The company invested $500,000 in upgrades to equipment

*www.usgbc.org/ShowFile.aspx?DocumentID=2058, accessed May 27, 2007.

efficiency, employee practices (such as recycling), and operations (landscaping practices), resulting in $610,000 in annual savings. Using an 8 percent cap rate, the annual savings should increase the building's value by about $7.5 million, or about $7.90 per square foot. In 2002, the building received an ENERGY STAR rating of 96 (out of a possible 100).*

CASE STUDY: CALIFORNIA DEPARTMENT OF EDUCATION, SACRAMENTO, CALIFORNIA

Also in Sacramento, in 2005, the California Department of Education certified its 600,000-square-foot headquarters building to the LEED-EB Platinum standard (Fig. 5.6); this was the first instance of a building previously LEED-NC certified (in this case, at the Gold level) going through the LEED-EB process. This was also the

Figure 5.6 The state Department of Education is a mammoth building across from the State Capitol in Sacramento. Achieving LEED-EB Platinum in 2005 engaged a large task force and resulted in energy savings of more than $600,000 annually and more than 600,000 gallons of annual water savings. *Photo by Erhard Pfeiffer, courtesy of Fentress Architects.*

*Ibid.

first state-owned building in the country to achieve a LEED-EB rating at any level. The initial "business case" stemmed from a 2004 executive order from Governor Arnold Schwarzenegger that all new state buildings had to meet a LEED-NC Silver standard. Here's how the team at the State Department of General Services (DGS) accomplished this significant undertaking, according to Theresa Townsend, at the time a senior architect with the DGS Green Team.*

> The Department of General Services decided that we would first aim for Platinum on our greenest building, the LEED-Gold-certified Department of Education, because we felt that we would be able to challenge ourselves and learn from each and every one of the credits that we attained. So by doing more, we're able to know what we're able to do on all of our other buildings and how the whole rating system works.
>
> Because this was the state's first LEED-EB project, there was a huge learning curve and initially it was very overwhelming. Probably the hardest was finding a team member to tackle each credit. It was a team of about 40 people that put all of this together. We had custodians, building engineers, managers of the building, energy specialists, architects, CAD drafters and industrial hygienists. We have a lot of very good people working for us, so most of the costs can be attributed to staff time. A very small amount was attributed to upgrading the building because it was a Gold-rated building to begin with. We are working now (2009) on some of our older buildings and I'm sure the cost will change substantially.

The California Department of Education project offered incentives to employees to promote the use of mass transit, bicycles, and alternative fuel vehicles; increased thermal comfort through an underfloor air distribution system; and instituted green cleaning practices to help improve indoor air quality. The building is now cleaned from top to bottom using a systems approach, with each team member having a specialty rather than having crews clean floor by floor. Green Seal chemicals and other environmentally friendly cleaning tools, such as microfiber cloths and low-decibel power equipment, are used.

To help communicate their green efforts, the State of California developed the *California Best Practices Manual: Better Building Management for a Better Tomorrow*, a set of guidelines that detail how facilities managed to attain the LEED-EB Platinum certification.† Following the successful certification of the state Department of Education building, the DGS led an effort to certify four additional large office buildings for the state Department of Public Health & Health Care Services, all at the Gold level.‡

---

*Interview with Theresa Townsend, April 2009.

†www.green.ca.gov/GreenBuildings/bbbtmanual.htm, accessed May 23, 2009.

‡http://www.cleanlink.com/hs/article/LEED-Facilities-Going-For-The-Gold-And-Even-Platinum—9367, accessed May 23, 2009.

Summary

The business case for green buildings does not rest solely on tangible economic benefits, but on many other tangible and intangible factors, including productivity and health gains, public relations, employee relations, access to financing, and building reputational capital as a sustainable enterprise. Economic benefits are of course important, and this chapter presents the economic and financial justifications for greening existing buildings. The prospect of future regulation of carbon emissions will also drive the business case for energy-efficiency building upgrades on the part of large building owners, operators, and managers. In the next few years, greening existing buildings will become an even more important part of a company's sustainability initiatives, demonstrating that it does, in fact, "walk the talk."

6

COSTS OF GREENING EXISTING BUILDINGS

Costs are the single most important factor in the building owner's world. The reason is simple: Costs are "hard" because they are real and occur in the present (and in the short term, revenues are fixed, so extra costs reduce profits), whereas benefits such as projected energy savings, water savings and productivity gains, though significant in the long run, are "soft" because they are speculative, may accrue to others and always occur in the future. Therefore, a cost-benefit analysis at the beginning of each LEED-EB project is crucially important to convince building owners, managers, and other stakeholders to proceed with the LEED certification effort.

The biggest barrier to greening existing buildings is the perception that they cost more to the owners than they deliver in the way of benefits. The World Business Council for Sustainable Development reported this widespread perception in an international survey in the summer of 2007. More than 1400 respondents in a global survey estimated the additional cost of building new green buildings at 17 percent above conventional construction; more than triple the actual cost difference of about 5 percent of original budget. At the same time, survey respondents put greenhouse gas emissions from buildings at 19 percent of world total, while the actual number of 40 percent of total emissions is more than twice the amount, counting emissions from both residential and nonresidential buildings.

Even experienced building developers and owners can have false perceptions about a business they know so well. Therefore, the only thing that will overcome this perception of high costs for greening existing buildings is to demonstrate that the process can produce higher performing buildings at reasonable costs, with attendant benefits. The other factor, of course, is to see other people doing it and not wanting to get stuck with a noncompetitive real estate portfolio. For most real estate developers and owners, the best course of action is to understand the value created by greening existing buildings, along a full range of value attributes, including but not limited to strictly financial returns.

Many of the interviews conducted for this book verified that people who "know what they are doing" can in fact get these net benefits, no matter what process they follow. Craig Sheehy is a LEED-EB consultant who directed the first LEED-EB Platinum certification, the California Environmental Protection Agency (Cal/EPA) building in Sacramento, California, while at Thomas Properties Group. He says:*

> What really blew me away, when I got into this, was how easy it is for a well-managed Class A building to achieve LEED-EB. So much of it is policy and procedure-driven that once we get the policies and procedures in place, it's quite easy. You have to have 34 points to achieve certification.† If you have a building with a 75 ENERGY STAR score, I can get you 29 of those points at no or absolutely low cost. Then to put up the balance is some serious payback. In LEED, it's all about the points. It is figuring out what points provide the biggest economic payback or return for that owner and that's what we go after.
>
> I'll perform an analysis on a building and we'll say 40 points is the minimum goal. I'll say, "Here's what it costs to get you certified. If you want to go to Silver, if you spend this money on a recommissioning of the building, which is an energy audit, implementing the low-cost measures that the audit finds and then putting in a continuous commissioning plan, that will pick up six points in LEED.
>
> I have not found a building yet that has been greater than a 14-month payback. Most of mine are in the 9- and 10-month payback range. That's a pretty fast payback. What I'm finding also is because utilities are starting to find out that these retrocommissioning energy audits provide such wonderful savings, they are offering some pretty good rebates. In California, the large regulated (investor-owned) utilities will do the audit for free.

Case Study: University of California Office of the President, Oakland

The University of California Office of the President (UCOP) is located in a 12-story office building (the "Franklin Building") in downtown Oakland, California, completed in 2007. The LEED-EB project received a Silver certification by achieving 40 out of 85 possible points.‡ According to the university, the certification process required 1500 staff and consultant hours and a net cash outlay of only $37,200. The project committed to annual costs of $1300 in operational changes that maintain the building with greater sensitivity to environmental and human health concerns. Annual savings of $30,700 so far have paid for both the initial cost and the small ongoing maintenance costs.

The bulk of the out-of-pocket costs were paid for building commissioning, a prerequisite in LEED-EB version 2.0; funds for commissioning were offset by a grant

*Interview with Craig Sheehy, April 2009.

†Sheehy is referring to LEED-EBOM 2008; in LEED-EBOM 2009, 40 points are required for certification.

‡Case study written by Trista Little, available from Matthew.StClair@ucop.edu.

from the Pacific Gas & Electric Company. The project resulted in an estimated reduction of 5 percent of electricity use and 3 percent of gas consumption.* By replacing faucet aerators in restrooms with 0.5-gallon-per-minute models, water consumption decreased by 20 percent. At a cost of about $300 per month, the Franklin Building (Fig. 6.1) increased its purchase of Green-e certified green power from 17 to 45 percent in support of the clean energy and climate protection goals in the university's policy on sustainable practices.

Figure 6.1 The UCOP project shows the potential for greening a fairly ordinary office building, cutting water use 20 percent, with strong leadership from a key tenant. *Photo courtesy of Trista Little, UC Office of the President.*

*Presentation by Jubilee Daniels to the "UC/CSU/CCC Sustainability Conference," June 25, 2007.

Other project highlights included:

- A new purchasing policy addresses the sustainable procurement of office paper, office equipment and supplies, and furniture. Sixty-eight percent of the building's total purchases now qualify as green under the criteria set by LEED-EB.
- All janitorial paper products meet the requirements of the U.S. EPA's Comprehensive Procurement Guidelines.
- Climate Neutral-certified carpet tiles, donated by Interface Flooring, were installed in all of the building's elevators.
- A new indoor air quality policy mandates low-VOC (volatile organic compound) levels for all adhesives, sealants, paints, coatings, carpet, carpet cushions, composite panels, and agrifiber products.
- A new composting program combined with a desk-side mini-bin system that emphasizes recycling has raised the building's waste diversion rate to just over 60 percent.

Cost Drivers for Greening Existing Buildings

What drives the costs of greening existing buildings, especially those project teams using the LEED-EB process? First let's look at factors relating to upgrade and renovation costs. Then we'll look at factors relating to LEED-EB process management. Table 6.1 shows the key LEED-EB project cost influencers, in the author's experience.

TABLE 6.1 KEY INFLUENCERS OF LEED-EB PROJECT COSTS*

TYPE OF DRIVER	RELATIVE INFLUENCE
Owner's experience with building retrofits	Medium
Team experience with LEED-EB projects	High
Level of LEED certification desired	Medium to high
Type of ownership	Medium
Team structure and process	Low to medium
Certification process and scope, including volume certification	Low to medium
LEED documentation difficulties	Low to medium
Consultant fees and internal time requirements	Medium to high

*Assessment of importance by the author.

DETAILED ANALYSIS OF COST DRIVERS

Let's now examine each of these drivers, in turn. While this discussion is general, it is based on the experience of many of the LEED project teams and consultants interviewed for this book, as well as the author's own experience in certifying LEED projects. It's important to note that costs can decrease over time both through individual team learning as well as information sharing among both LEED consultants and building owners.

Owner's experience with building renovations and retrofits Clearly, the single greatest factor in crafting a LEED-EB project is what has to be done to the building to bring it up to the LEED standard. Because LEED-EB requires a minimum ENERGY STAR score of 69 and a minimum water savings of 20 percent below a baseline based on current plumbing codes, these two items will drive costs the most. Many of the people we interviewed for this book mentioned that the easiest projects are relatively new buildings that already have an ENERGY STAR score of at least 70, meaning they require no physical upgrades to building envelope or heating, ventilating, and air-conditioning (HVAC) systems to meet LEED's minimum standards. Similarly, buildings constructed since 1993 already meet current plumbing codes and require relatively minor upgrades to meet the LEED water conservation prerequisite.

Team experience with LEED-EB projects Team experience with LEED-EB projects is obviously a critical factor for two reasons. First, teams with little or no experience may naturally assign a "risk premium" to their costs, especially true for consultants hoping to get paid for their "learning curve" by the first client. Second, teams with experience will have developed shortcuts, written standard policies, researched alternative approaches, and generally have learned to pick team members with a "get it done" attitude. Therefore, they don't need to assign a risk premium to their costs and are more likely to treat LEED-EB project requirements as "business as usual" rather than an added burden to their continuing operations.

LEED certification level The level of LEED-EB certification desired or required is clearly a cost driver. As one seeks higher levels of LEED certification, even with a newer building, one is more likely to add higher cost elements such as green roofs, photovoltaics, significant energy upgrades, and major water conservation measures. Additionally, it may be harder to confine the performance period to just three months, since rewriting and implementing a wide variety of company or organizational policies cannot often be accomplished quickly.

Type of ownership It should be easier and cheaper for an owner-occupied building to achieve certification (certainly at higher levels) than for a multitenant building, because of the requirement to gather lots of information on such things as purchasing and waste management policies of the tenants, as well as to survey occupants, get permission for energy upgrades in tenant spaces, and in general to "herd cats" with multiple tenants. However, when one very large property management company such as

CB Richard Ellis decides to pursue LEED-EB on a wholesale basis, it's likely that they will quickly learn how to get these projects done without significant cost premiums, compared with a single-tenant/owner-occupied building or facility. Indeed, some single-tenant buildings may in fact have similar difficulties in getting tenant cooperation as multitenant buildings; these might include academic buildings and very large corporate office buildings, for example.

Team structure and process As I will show later in the book, the approach taken by the team in organizing the process is critical to getting the job done, both in terms of time and cost. The more consultants there are on a team, the higher the costs, since coordination issues and high-level expertise contribute to these costs. This is particularly true for projects that need recommissioning and energy performance upgrades to an HVAC system or building envelope to achieve an ENERGY STAR score of at least 69 or to meet minimum required ventilation standards.

However, a key element in team structure is the role of the internal process leader, that is, how effective that person is in organizing the internal team and how committed he or she is to the final outcome.

Certification process and scope The LEED certification process has frequently been accused of being overly cumbersome and burdensome, in terms of the opacity of the process and the requirements for documentation. In my experience, however, it's only the first project that confuses most people. Once you've seen the process from beginning to end, it begins to make more sense and becomes easier to navigate. Experienced LEED-EB consultants interviewed for this book have developed methods for streamlining project delivery. After all, if a building's operations actually satisfy the LEED-EB requirements in daily practice, it shouldn't be that hard to document compliance.

It's also important to do most of the key thinking during early project meetings and to use whole-building energy models (as was done for the Empire State Building upgrade, profiled in Chap. 1) and also the ENERGY STAR Portfolio Manager to get better and earlier decisions on energy upgrades.

LEED documentation difficulties Many teams experience difficulties the first time through a LEED project with the very specific documentation requirements and the amount of effort needed by all responsible parties to get the documentation right from the beginning, or else face an extensive review cycle if documentation is rejected by the LEED reviewers. For LEED, getting all documentation for the same quarterly performance period is often the first hurdle. The second hurdle is often the effort required to put information that the organization might already have (energy costs, water costs, waste management costs, etc.) into a LEED-acceptable format. And of course, acquiring the necessary information can be daunting (think of finding out about everyone's commuting arrangements in a large multitenant building). Getting documentation right the first time is certainly the way to keep costs under control. This often requires the use of outside consultants, which adds to project costs.

Elaine Aye is a LEED consultant with nearly 10 years of LEED experience. From her vantage point, she points out the need to use all available resources, such as vendors and possibly even summer interns to get the process done.*

> Most facility staffs run pretty lean, so we're encouraging building owners to work with their vendors and bring them to the table—it may be their janitorial company, landscape maintenance company, it may be their service contractors or a building service engineer that supported their building—and help in the process. We also encourage hiring a summer intern who can come in and manage the paperwork and coordination on getting some of the systems in place.

Consultant fees and internal time requirements The requirement for LEED documentation could cost between $25,000 and $50,000 for team coordination and LEED project management services. Whether it's performed in-house at a business, government agency or nonprofit organization, or done mostly with outside consultants, there is a higher level of effort required to coordinate all the project team members and to keep the LEED aspects of the project on track during the performance period. Often, teams might find themselves spending considerable staff time trying to understand and document things that are relatively simple to do with the aid of an outside consultant.

On this subject, Craig Sheehy describes how he works to minimize both the in-house time commitment and the outside consultant fees:†

> Our process is a 20-week process. Property managers and engineers don't have enough hours in the day to do their job right now. They get a little nervous and a little scared when they're asked to get the building LEED-certified. My goal is to eliminate as much work as possible for the building team; our staff will handle everything that we possibly can. There are only about seven to ten points that we'll go after in a project that we can't get without the staff's help. We set up the project so they will be working on LEED only one to two hours per week during the 20-week project. They're never overwhelmed. They're never buried and I think the greatest part of it is, that in all the buildings we've gone through, the property manager and engineer still like us when we're done.

From Sheehy's comments, one can conclude that if building owners and managers want high-performance green buildings, they should be willing to pay what it costs to engage the very best consultants and engineers to accomplish that result. By the same token, the owners should push the certification team to figure out how to reduce total LEED-certification project costs with their recommendations. It's obviously worth paying higher consultant fees if you can get a less expensive overall result.

---

*Interview with Elaine Aye, April 2009.

†Interview with Craig Sheehy, op. cit.

Project Profile: 717 Texas, Houston

This building was developed and is owned, managed, and leased by Hines, a large international firm with a history of strong ties to both the ENERGY STAR and LEED programs.* This project received a LEED-EBOM Silver rating in early 2009, achieving 46 of a possible 92 points. 717 Texas is a 33-story, 696,000-square-foot multitenant office building.†

In 2002, the building participated in the LEED for Core & Shell pilot program and has been regarded as perhaps the most sustainable office property in downtown Houston. Designed with sustainability in mind, many green features reduce energy consumption and promote a healthier tenant work environment, including: a green and reflective roof on the parking garage, which reduces ambient temperatures; electrostatically purified outside air and a green cleaning program, which provide superior indoor air quality; filtered and purified drinking water to reduce the use of bottled water; low-flow restroom fixtures for water conservation; and a recycling program that captures 75 tons of material annually, more than 50 percent of all waste generated.

Shown in Fig. 6.2, 717 Texas is also ENERGY STAR-qualified; with a score of 86, the building performs 36 percent better than if it were operating at the national average for energy performance. This building illustrates Hines' strong commitment to high-performing buildings. The firm has more than 98 projects, representing approximately 65 million square feet that have been certified, precertified, or registered under the various LEED programs. Hines has also received an ENERGY STAR label for more than 130 buildings, representing approximately 76 million square feet.‡

According to Adam Rose, Hines' property manager at 717 Texas, the LEED process was fairly time-consuming, but was facilitated by an outside consultant:§

> I would probably use a consultant again because there's a lot of information to manage and the details of the rating systems change regularly. While we definitely have the knowledge base to do it on our own, especially with the help of our corporate LEED team, it is still beneficial to have a consultant to help with some credits that are more technical or beyond the regular scope of our work at the property. Our staff put about 800 to 900 hours into this process. So, there's absolutely a value in your time to hiring a consultant.

ADDITIONAL COST CONSIDERATIONS

There are other potentially significant factors that determine what a LEED-EB project will cost, on a "dollars per square foot" or "dollars per project" basis. Knowing what to budget is certainly important, since the money for the process has to come from somewhere. Unlike a new construction certification, the LEED-EB project budget cannot

*Personal communication, Clayton Ulrich, senior vice president, Hines, May, 2009.

†www.hines.com/press/releases/2-6-09.aspx, accessed May 20, 2009.

‡Ibid.

§Interview with Adam Rose, April 2009.

Figure 6.2 **717 Texas shows what can be done with an office building that has "good bones," and a committed owner, cutting energy use by 36 percent.** *Photo courtesy of Joe Aker, Aker/Zvonkovic Photography.*

be buried in a much higher project cost. While the benefits shown in Chap. 5 might be there, they are still hard to estimate without going through the process, particularly in those projects in newer buildings that involve no upgrades to the HVAC system. When estimating a LEED project, these factors often determine the final project budget. Some of them are quite significant but may have nothing to do with the level of LEED certification or energy performance sought.

Timing of the project A significant determinant of cost is when the project is started. Since the commercial new building construction sector has slowed down dramatically in 2008 and 2009, general and specialty contractors will likely be more eager for work, and prices for building upgrades should come down. Continuing deflation of construction materials costs in 2009 and 2010 may also be a factor in moderating costs for building upgrades.*

Project size One would expect smaller buildings to have a higher cost premium for LEED certification, because certain costs of LEED (e.g., documentation assistance and building recommissioning, to pick just two) have fixed-cost elements independent of project size that will add to the cost per square foot. Above a certain size, perhaps 50,000 square feet, this "size cost premium" starts to disappear. Let's say such a project has a budget of $0.5 million and that the LEED-EB "fixed cost" items for a Silver certification total $75,000. That would represent cost premium of 15 percent, generally tolerable for most projects, assuming there are not too many other net cost additions for achieving higher energy efficiency. At $1.0 million, the fixed cost premium would be less than 10 percent and at $1.6 million, less than 5 percent.

Feasibility of LEED-EB measures Another cost factor is the feasibility of LEED-EB actions. For example, in most urban areas, recycling a higher percentage building waste is typically a no-cost item for the LEED-EB project. However, for projects in rural areas, there may be fewer waste recycling opportunities. The same might hold true for many of the materials and resources credits that might rely on purchasing policies, in case there is no local supply chain for recycled content materials or certified wood products. Adam Rose is property manager at the 717 Texas building in Houston for Hines. He spoke of the practical difficulty of implementing some LEED measures at this property.

> Another hurdle is that a lot of vendors aren't as familiar with some of the processes and products. There were green products but our distributors weren't as familiar with them, what we needed and how they needed to be used. So we would find a product and then we would have to teach our cleaning staff what to do with it.

Controlling Costs in LEED Projects

From a number of the participants interviewed, the following "lessons learned" on controlling costs can be gleaned:

- Have a clear certification goal from the outset; if you want Gold or Platinum, make it clear upfront and provide sufficient budget to reach that goal.
- Make sure the project team includes all disciplines and specialties that will be needed in the certification process.

*"Construction Costs Post Rare Quarterly Decline," Engineering News-Record, June 29, 2009, p. 21, www.enr.com.

- Incorporate all relevant LEED-EB elements in the team's considerations from the beginning, starting with the prerequisites but also including key policies and aspects such as recommissioning that will affect many other decisions.
- Have centralized management of the LEED certification project, with clear responsibilities assigned to each team member.
- Team members should understand green building issues and practices; there may be a need for an introductory workshop or two to explain the rationale behind all of the LEED requirements.
- Obtain sufficient technical information to make informed decisions; make sure that the people who have this information are integrated from the beginning into the project team.
- Provide sufficient upfront time and funding for any relevant technical studies and projects such as recommissioning.
- Always insist on life cycle costing of proposed green investments; that will make it easier for upper management to support the project.

I will return to these points in several places in this chapter, since each project team has to address the challenge of identifying greening costs (and benefits) and justifying them to upper management.

Many of the green building measures that give a building its greatest long-term value—for example, on-site energy production, on-site stormwater management and water recycling, green roofs, daylighting, and natural ventilation—often require an investment cost. While it is often possible to get a LEED-EB-certified (and sometimes LEED-EB-Silver) building at no additional cost, cost increments often accrue as project teams try to make an existing building more sustainable. This is especially true when the building owner or manager wants to showcase their green building with more expensive (but visible) measures such as green roofs or photovoltaics for on-site power production, or where there is a strong commitment to using specific green materials in building renovations.

Project Profile: Cal Poly Faculty Offices East, San Luis Obispo, California

Located on California's central coast, about 190 miles north of Los Angeles, California Polytechnic State University is one of 23 campuses in the California State University System. Cheryl Mollan is the project manager for the campus facility department and was the LEED-EB project administrator for the Silver certification of the 28,000-square-foot Faculty Offices East building (constructed in 1991) under LEED-EB version 2.0, completed in October 2008 (Fig. 6.3).*

*Information provided by Cheryl Mollan, Facilities Department, Cal Poly State University, March 2009.

Figure 6.3 **The Cal Poly Faculty Offices East reduced water use by more than 65 percent inside and outside the building, achieving Silver certification with a wide-ranging program of green measures.** *Photo by Cheryl Mollan, courtesy of California Polytechnic State University.*

For this project, building upgrades included the following:

- Replacement of restroom fixtures with low-flow fixtures, including urinals, toilets, and faucet aerators and sensors, resulting in 73 percent water reduction.
- Installation of countertops with 85 percent recycled glass content.
- A 65 percent reduction in water use for plantings through using mulch produced by landscape waste on campus, introducing native plants, and employing high-efficiency landscape irrigation controllers.
- New electric hand dryers reduced the amount of waste generated from paper products (installed on second floor restrooms).
- Sensor-activated low-level exterior bollards and bi-level stairwell lighting are anticipated to save 75 percent of the lighting energy used by conventional T8 fluorescent lamps.
- Carpet with 40 percent recycled content was installed in the conference room.
- "Smokeless" urns (located at designated smoking areas) reduce airborne contaminants by extinguishing a cigarette immediately upon disposal and also reduce

ground-keeping labor as the urns are emptied less frequently. The urns are fabricated with 65 percent recycled content.

- Installation of window tinting was used as a means to reduce solar heat gain.
- Commissioning of all HVAC systems was combined with associated repair or replacements for optimal energy efficiency.
- Recycling 75 percent of construction waste.
- Green products account for 90 percent of cleaning product purchases.
- Paints and sealants used in restroom upgrades contain zero VOCs.

For this project, Cal Poly staff provided the majority of building upgrades and all certification documentation, serving to create an opportunity for professional development and to secure significant savings in the LEED-EB certification process. This project also was used as a pilot project for testing new sustainable products for future use in other campus buildings.

Summary of LEED-EB Project Cost Influences

Chapter 5 discussed the many business benefits of greening existing buildings, and this chapter has made the case that costs are real, occur first, and must be justified to various stakeholders. Benefits are generally long term, and costs are immediate, so many people tend to shy away from anything that will add costs, no matter how large the potential benefits.

Table 6.2 shows some of the decision elements of greening existing buildings that may influence total project cost. From this table of "cost influencers," you can see that there is no right answer to the question: "How much does greening an existing building cost?" To anyone who asks, I often say that the definitive answer to this question is simple—it depends!

Higher levels of sustainable achievement (e.g., LEED-EB Silver, Gold, or Platinum) may involve some incremental capital costs and may also incur further soft costs for additional design, analysis, engineering, energy modeling, building recommissioning, and certification. For some projects, additional professional services, for example—including energy modeling, building commissioning, added design analyses, and the documentation process—can add 0.5 to 1.5 percent to cost, depending on a project's size.

In terms of "hard costs," consultant Barry Giles says there are at least three showstoppers in the LEED-EBOM system:*

> One of those is Indoor Environmental Quality Prerequisite 1, which is testing the fresh air delivery into the building. That is posing immense financial difficulties [in some cases].

---

*Interview with Barry Giles, April 2009.

TABLE 6.2 COST INFLUENCERS FOR GREEN BUILDING PROJECTS

COST INFLUENCER	POSSIBLE COST INCREASES
1. Level of LEED certification sought	Zero for LEED-EB certified to 1 to 2% for LEED Silver, up to 5% for LEED Gold
2. Project type	With certain project types, such as labs, it can be costly to change established systems; systems for office buildings are easier to change
3. LEED-EB experience	Every organization has a "learning curve" for LEED-EB; costs decrease as teams learn more about the process
4. Specific "green" technologies employed	Photovoltaics and green roofs are going to add cost, no matter what; it's possible to get a LEED-EB Gold building without them
5. Lack of priorities for green measures	Each team member considers strategies in isolation, in the absence of clear direction from the owner, resulting in higher costs overall and less systems integration
6. Geographic location and climate	Climate can make certain levels of LEED certification harder and costlier for project types such as labs and even office buildings

> We've gone from LEED-EB version 2.0 requiring a commissioning process at a cost of 25 cents to $1 per square foot, to this air-handling requirement that is causing just as much financial difficulty because it requires every single air handler to be tested. In some buildings, that can mean 170 air handlers need to be tested. That costs money and as a prerequisite, it has to be done. Clients are balking about it a bit, especially in this recession.
>
> The other showstoppers are ENERGY STAR—you've got to get an ENERGY STAR score of 69—and the water prerequisite, the minimum water efficiency [savings of 20%]. The snag is that in big buildings it takes time to do, it's very detailed and if you are not successful, you have got to spend money to raise your ENERGY STAR score. There's no way around it. If you've got to replace china (toilets, etc.), it's expensive. It's a stall point for a client.

Paul Goldsmith is a sustainable operations champion with Harley Ellis Devereaux who has worked on a number of LEED-EB projects. He concurs with Giles, saying:*

> The two hard ones are energy and ventilation on older buildings. A team may have to stop for a while and look at certain improvements to ensure that they're able to achieve the ENERGY STAR and ventilation requirements, especially in older buildings.

*Interview with Paul Goldsmith, April 2009.

I've been involved in a good number of buildings where we've hit that wall, they've stopped for a while and now with the impact of the economy they're holding off for now. They felt that they had an efficient building and they thought they had good ventilation, but in reality they are still not up to par.

Green Building Cost Studies

Given the high level of interest in the costs of green, it's surprising that there are so few rigorous studies of the cost of "green," especially those comparing similar projects. Here's one study that points the way toward a better understanding of the costs of greening existing buildings using the LEED-EB process.

THE LEONARDO ACADEMY STUDY

In 2008, the Leonardo Academy, the original support consultant for the LEED-EB program, issued a report on the costs and benefits of LEED-EB. Table 6.3 shows the results of that study. However, one should keep in mind that these costs were for projects certified during 2006 and 2007 and reflect the project teams' experience with the LEED-EB version 2.0 system, one that had more prerequisites and is no longer in use. Note that the table only includes LEED-EB Gold and Platinum certified projects, four of each, so that the data should be used only as a guideline for estimating your own LEED-EB costs.

Consultant Elaine Aye has focused specifically on LEED-EB for several years. As for in-house costs, she says:*

TABLE 6.3 MEDIAN LEED-EB IMPLEMENTATION AND CERTIFICATION COSTS (PER SQUARE FOOT)*

COST ELEMENT	GOLD PROJECTS ($n = 4$)	PLATINUM PROJECTS ($n = 4$)
Staff hours (internal labor)	10 per 1,000 square feet	5 per 1,000 square feet
Staff costs (internal labor)	$0.40	$0.33
Consulting costs	$0.27	$0.79
Total soft costs (w/labor)	$1.47	$2.05
Total hard costs	$1.77	$0.11
Total all costs	$3.24	$2.16

*Leonardo Academy, "The Economics of LEED for Existing Buildings, for Individual Buildings, 2008 Edition," April 21, 2008, pp. 5–6.

*Interview with Elaine Aye, op. cit.

We've seen clients spend as little as 200 hours in implementing the EB process, to some spending over 1000 hours. That's just overkill on their part—weekly meetings with six people in the meeting, for example. It really varies as far as in-house costs are concerned.

Soft costs for LEED-EB projects One thing is certain: Soft costs need to be taken into account, especially for projects aiming at higher levels of certification. Many project managers do not consider these costs especially onerous, but some do. Table 6.4 shows the source of some of the potential soft costs. Some of these costs should be considered essential to good project design and execution, specifically building commissioning and energy modeling, while others are more clearly associated with the LEED certification effort.

Some consultants provide much lower estimates, especially for larger buildings. Craig Sheehy says, "We're averaging about 21 cents per square foot, all-in costs for LEED-EB—that's registration, certification, consultation, any retrofits that took place or any energy audits/retrocommissioning that took place."* Note that there are certain fixed costs with any LEED-EB certification project, so that larger buildings, those for which Sheehy tends to provide his service, will have lower unit costs than smaller projects.

TABLE 6.4 "SOFT COSTS" OF LEED-EB CERTIFICATION, 2009*

ELEMENT	COST RANGE	REQUIRED IN LEED-EBOM?
1. Recommissioning	$0.40 to $1.00 per square foot	No
2. Energy modeling	$15,000 to $30,000	No; use EPA's Portfolio Manager tool instead
3. LEED documentation	$25,000 to $75,000	Yes; depends on complexity of project, team experience, and level of certification
4. Project team meetings	$10,000 to $50,000	Not required, but necessary
5. Emissions reporting	$5,000 to $15,000	No
6. Daylighting modeling	$3,000 to $10,000	No (some utilities offer this as a free service)
7. Measurement and verification plan	$10,000 to $30,000	No

*Leonardo Academy, op.cit.

*Interview with Craig Sheehy, op. cit.

Presenting Costs and Benefits: "Payback" versus "Return on Investment"

I've often observed that many engineers (and to some extent, contractors and facility or building managers) need to improve their presentation of the economic benefits of LEED to higher-level decision-makers who prefer to speak the language of finance not building operations. A good example occurs in the world of energy savings. For example, suppose it costs $300,000 extra to secure annual energy savings of $100,000. This is a typical circumstance. An engineer might say that this measure has a "three-year payback" ($300,000/$100,000) in terms of how long it takes to recover the initial investment from the annual energy cost savings. A more business-oriented person would say that this measure has a "33 percent, inflation-protected return on investment," because payback is measured in terms of today's energy costs, which are quite likely to increase faster than inflation in the future. Which approach sounds more inviting—waiting three years just to get your money back, or making a very high-return investment? Both use the same data, but one is more likely to get approval, don't you think?

Let's consider the same situation in the world of commercial real estate. Commercial properties are usually valued as a multiple of "net operating income," typically determined by dividing the income by a capitalization rate expressed as a percentage (think of how a corporate bond is valued; it's the same approach). Reducing annual energy costs by $100,000, using a typical "cap rate" of 8 percent, would yield an incremental increase in value of $1.25 million ($100,000/0.08). So the same investment in energy efficiency (in a commercial situation) would create greater than a 400 percent immediate return on investment ($1,250,000/$300,000)! Quite a difference, don't you think, between a "33 percent return on investment" (which of course sounds pretty good) and a "400 percent immediate increase in value!" If LEED-EB proponents would take the time to learn how their clients make and talk about money, it would make it far easier to "sell" green building upgrades for many projects.

Architect Clark Brockman is a strong proponent of learning how to talk the language of finance and to be realistic in presenting the long-term value of greening a building. He says:*

> With LEED-EB, it seems that some of our clients want whatever measure they invest in to provide a simple payback in two years or less, but then when you ask them how long they expect to be in the building they say, "15 or 20 years, or maybe forever." At this point we (the green building community) need to change the discussion away from "simple payback" to return-on-investment (ROI) by asking, "Isn't a seven-year investment that cuts your utility bills by $100,000 a year for the life of the building a good investment?" The design community has to be able to shift the market [mindset] by shifting the discussion from "first-cost" to "value-over-time."

*Interview with Clark Brockman, April 2009.

Summary

Because of the relatively recent growth in the LEED-EB system, information on the costs of achieving various levels of certification is still relatively limited, so it's hard to budget such projects. Nonetheless, by considering upfront many of the factors that influence cost, accessing the experience of other teams, using experienced consultants, and understanding the potential for realizing significant benefits, many projects have been able to control costs while moving ahead with greening existing buildings.

7

MEETING THE ENERGY RETROFIT CHALLENGE

In previous chapters, I've reviewed the forces leading to a renewed interest in greening existing buildings, introduced you to the LEED rating system, outlined the business case for greening existing buildings, and presented some of the key cost considerations. Now it's time to get down to the details of how to accomplish these valuable upgrades. In this chapter, I'll focus on energy-efficient retrofits, since those constitute more than one-third of the LEED-EBOM 2009 rating system points and offer the most tangible return on investment. Thus, energy improvement measures are likely to be the first things undertaken in most projects. In fact, one can accumulate almost enough LEED points with energy efficiency upgrade measures to certify a project, provided that the project meets all the prerequisites outlined in Table 4.3.

LEED-EBOM Energy and Atmosphere Credits

LEED provides the following points for specific achievement levels in six categories related to energy savings, renewable energy use, and atmospheric protection. Table 7.1 lists the credits and associated LEED points. One point of clarification about nomenclature: in LEED, a "credit" is a specific category of environmental concern such as energy efficiency or renewable energy use. A "point" is awarded for a specific achievement in a particular credit category. Some credits have only one point; some have many, as you will see. One further clarification: most credits are "performance" credits, that is, they specify a desired outcome, such as an ENERGY STAR score, but don't tell you exactly how to achieve it. A few credits represent "prescriptive" approaches that tell you exactly what you must do to secure the relevant LEED points. Since LEED is a rating system based on "best practices," there is a blend of both approaches.

TABLE 7.1 LEED-EBOM ENERGY AND ATMOSPHERE (EA) CREDITS

CREDIT	POINTS
Prerequisite 1—Energy Efficiency Best Management Practices (BMP)	None; requires ASHRAE Level I "Walk-Through" Energy Analysis
Prerequisite 2—Minimum Energy Performance (ENERGY STAR score)	None; minimum ENERGY STAR score of 69 required
Prerequisite 3—Fundamental Refrigerant Management (No CFCs)	None
1. Optimize energy efficiency performance (ENERGY STAR score)	1 to 18 points, based on ENERGY STAR score of 71 to 95, with one additional point awarded for each increment on a sliding scale
2.1 Commissioning (Cx)—investigation and analysis	2
2.2 Commissioning—implementation	2
2.3 Commissioning—ongoing Cx	2
3.1 Performance Measurement—building automation system in place with preventive maintenance program	1
3.2 Performance measurement—system-level metering; meter 40 or 80% of total expected annual energy consumption	1 to 2
4. On-site and off-site renewable energy	1 to 6, based on contribution to total energy use, ranging from 3 to 12% for on-site power generation and 25 to 100% for off-site purchased power
5. Enhanced refrigerant management—reduction of HCFC refrigerants	1
6. Emissions reduction reporting—using a recognized third-party program	1

Energy and Atmosphere category points total 35 in all. Compared with LEED EBOM 2008, the 2009 version raises the ENERGY STAR score and adds building commissioning. Because building commissioning (commonly abbreviated as Cx) is so fundamental to improving energy performance, I'm going to explain it in more detail. In all other LEED systems, commissioning is a prerequisite, with no points attached, but in LEED-EBOM it was wisely given up to six points to emphasize how important Cx is to improving energy performance.

Building Commissioning*

In LEED-EBOM's Energy and Atmosphere (EA) credit 2.1, the first step is to develop an understanding of how the building's major energy-using systems are operating, then to specify operational changes that will optimize energy performance, and finally to craft a plan for achieving the targeted energy savings with capital improvements and operating changes. The first step is usually to "break down" the energy use into major categories such as lighting, cooling, heating, hot water, etc. (Figure 7.1 shows primary energy use by various energy end-uses, which includes the fact that electricity is the primary fuel for commercial buildings and is only about 30 percent efficient, including energy losses at both the generation and transmission stages.)†

Also known as an "energy survey and analysis," one may follow the procedures of an ASHRAE Level II Energy Audit.‡ (A walk-through analysis, called a Level I by ASHRAE, was already required by Prerequisite 1). ASHRAE (the American Society of Heating, Refrigerating, and Air-Conditioning Engineers) is an important source for energy analysis and design standards in the United States and Canada. The International Energy Conservation Code (IECC), promulgated by the International Code Council

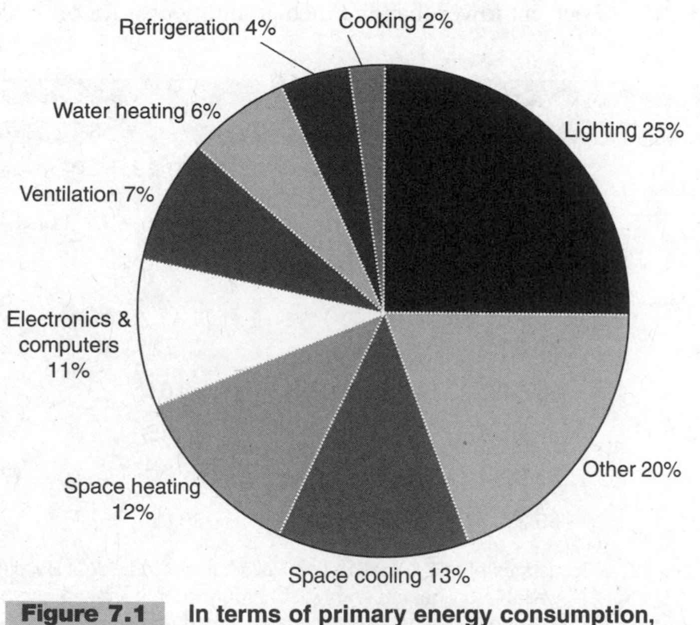

Figure 7.1 **In terms of primary energy consumption, lighting is often the single largest end use in commercial buildings, followed by cooling and heating.**

*LEED Reference Guide, pp. 169–190.

†www.ferc.gov/eventcalendar/.../20070216090203-2-14-07mediab.pdf, accessed July 7, 2009.

‡ASHRAE, 2004, *Procedures for Commercial Building Energy Audits*, RP-669, SP-56, available from www.ashrae.org.

incorporates ASHRAE 90.1-2007 for commercial building standards. States must adopt IECC as a condition for receiving the 2009 ARRA stimulus law funding.*

The set of procedures contained in EA credits 2.1, 2.2, and 2.3 is often called "retro-commissioning" or "recommissioning" or even "continuous commissioning," although some experts may feel there are subtle differences between the terms. This book will use "retro-Cx" as the main term. The next credit, EA credit 2.2 is for actual implementation of the recommendations in EA credit 2.1, specifically no-cost or low-cost operational improvements, along with developing a specific budget for higher-cost retrofits or upgrades. The building's operating plan should also be updated, and building management staff needs to be trained in the new building operations plan.

Finally, EA credit 2.3 awards two points for implementing an ongoing commissioning plan that integrates "planning, system testing, performance verification, corrective action response, ongoing measurement and documentation."† Anyone who has been married for a long period will recognize these steps as essential to a successful relationship. It's the same with buildings: nothing is ever static, everything must be continually reviewed, and performance upgrades are a regular feature of building operations.

How important is commissioning? Since utility costs can represent 30 percent of building operating costs in conventional buildings, as shown in Table 7.2, it makes sense to try to reduce them. Table 7.2 shows that LEED-EB certified buildings have a median utility cost more than 30 percent lower than the median utility cost for all buildings in

TABLE 7.2 MEDIAN BUILDING OPERATING EXPENSES ($ PER SQUARE FOOT)*

COST ELEMENT	BOMA 2007	PERCENTAGES	LEED EB	PERCENTAGES
Administrative	$0.99	14%	$0.78	13%
Cleaning	$1.14	16%	$1.24	20%
Repair/Maintenance	$1.52	22%	$1.17	19%
Roads/Grounds	$0.08	1%	$0.33	5%
Security	$0.54	8%	$0.04	1%
Utility	$2.11	30%	$1.45	24%
Other	$0.59	8%	$1.06	17%
Total expenses	**$6.97**		**$6.07**	

*Leonardo Academy, "The Economics of LEED for Existing Buildings, for Individual Buildings, 2008 Edition," April 21, 2008, p. 21. Available online: http://www.leonardoacademy.org/categoryblog/131-the-economics-of-leed-for-existing-buildings-for-individual-buildings-2008-edition.html, accessed July 7, 2009.

*ICC incorporated ASHRAE 90.1 in the 2009 IECC. The ASHRAE 90.1 reference maintains language contained in the 2006 IECC currently enforced by state and local jurisdictions. http://aec.ihs.com/news/energy-efficiency/2009/icc-energy-iecc-21209.htm, accessed July 9, 2009.

†Ibid., p. 185.

the United States. This is consistent with the experience of LEED-NC buildings that focus on energy reductions and require commissioning (and which show an average 30 percent reduction in energy costs),* so one can probably attribute most of the utility cost reduction to the Cx prerequisite found in LEED-EB version 1.0 and 2.0 systems.

EXISTING BUILDING COMMISSIONING

Paul McCown is senior project manager at national commissioning provider SSRCx,† with prior experience in LEED-EB certified projects. He sees existing building commissioning (Fig. 7.2) as different from new building Cx in the following way:‡

> From my experience, there are two ways to approach commissioning of an existing building. One is the traditional definition of retro-Cx, which is perhaps the ASHRAE definition of returning the building to [the original] plans and specs.

Figure 7.2 Retro-Cx involves a comprehensive look at all building systems, with the primary goal of improving energy performance. *Courtesy of SSRCx Facilities Commissioning.*

---

*Cathy Turner and Mark Frankel, 2008, *Energy Performance of LEED for New Construction Buildings*, Final Report, March 4, 2008, New Buildings Institute, www.newbuildings.org/measuredPerformance.htm, accessed May 27, 2009.

†www.ssrcx.com, accessed May 27, 2009.

‡Interview with Paul McCown, April 2009.

> The other approach is evaluating the building and energy consuming systems on the basis of optimizing those systems for energy efficiency. That's the primary objective, with a secondary emphasis on things like sensors out of calibration, [control] sequences that need to be reset, equipment that needs to be replaced and valves that are not operating properly (such as valves that are passing chilled water when they're supposed to be closed which creates a simultaneous heating and cooling effect.)

McCown and the team at SSRCx are advocates of the process of Continuous Commissioning®, a process developed and trademarked by Texas A&M University that goes beyond the "one-time" retro-Cx used in most LEED-EB projects:

> [At SSRCx], we've come to the conclusion that there's a set methodology for commissioning that encompasses training the owner and the facilities staff to help them understand how the systems in the buildings operate through the building management system and how the building management system works so they can be comfortable with it. An all-encompassing program is the most beneficial [way to achieve] optimal energy savings in the building and optimal equipment use.
>
> We are a provider of Continuous Commissioning, a quite comprehensive approach where you focus on energy and you get the other benefits of functioning systems secondarily, as a by-product. With Continuous Commissioning and the way that we approach it, we're actually trying to figure out new ways to harness energy efficiency from the existing systems.

Case Study: Caterpillar Financial Headquarters, Nashville, Tennessee

Caterpillar Financial Headquarters in Nashville, Tennessee, achieved a LEED-EB 2.0 Gold certification in 2009 (Fig. 7.3). Steve Zanolini is global facility manager for Caterpillar Financial and the internal leader in the LEED certification effort.* His experience in getting employee participation in some of the key LEED-EB initiatives, such as desk-side recycling, illustrates the importance of getting them personally involved in the sustainability program:

> We have a great company, but I cannot say that we had 100 percent buy-in from everyone in our move toward sustainability. Still, the vast majority of people appreciate the fact that we are a sustainable company. One of the obstacles [in the LEED-EB project] was explaining why this was necessary and proving that point. We did a lot of education in the building with help from our communications folks to help everyone understand, not only [what is] sustainability, but why sustainability [is important.] How does it benefit them? Over time, we got their buy-in.

*Interview with Steve Zanolini, May 2009.

Figure 7.3 **At Caterpillar Financial Headquarters in Nashville, Tennessee, the retro-Cx process helped the company identify and resolve a number of outstanding building performance issues affecting energy use and indoor air quality.** *Photo courtesy of Rob Congdon/Cat Financial.*

Every building has a history, and in this case, Caterpillar Financial was initially going to be only one of several tenants to occupy the building. Accordingly, the building was designed and built in a conventional manner. Zanolini says:

> Initially, the building was built to be a multitenant building. When we initially partnered with the owner and developer to occupy the building, we were only going to occupy about 50 percent of it. The motivation of the builder was not necessarily sustainability because they knew that they could pass on the cost of operating this building to their tenants, which is standard practice.

Zanolini's experience with LEED-EB certification led him to think about how this fits into the company's sustainability strategy.

> We had some obstacles (and we still do) in pursuing sustainability. I view the LEED process as one point in time, but I view sustainability as a forever, ongoing practice. I've established three goals for the building: zero-net electricity, zero-net potable water

and zero solid waste. These are long-term goals but it makes sense for the company because we have a long-term lease in the building. We're going to be here for a long time. It's going to be tough. And this is an existing building, which is going to make it much harder to accomplish those goals than if we were building a new building.

From initial occupancy in 2000, until the end of 2007, the building had been managed by an outside entity. The LEED-EB process was a good evaluation tool for Zanolini, in charge since the beginning of 2008, to look in detail at how the building was actually running.

When we assumed the management of the building, we faced a number of challenges when it came to the maintenance of the mechanical equipment of the building. We have variable-air-volume (VAV) air-mixing boxes in the ceilings of every floor, and there are anywhere from 18 to 20 per floor. Some of the boxes are fan-powered and some are not. Those with fans have a filter.

Since we're a company committed to Six Sigma [quality management]*—where facts and data are important—I wanted to assure that every one of those fan-powered boxes had a fresh and clean filter. I directed Trane Corporation to immediately check them. What we discovered was a lack of a good, systematic preventative maintenance program.

The retro-Cx process was conducted by SSRCx, and led Zanolini to further revelations about what was hindering building performance.

The commissioning process didn't actually start until the [LEED-EB] performance period. We had our engineering consultant SSR partner with Trane Company, and we had every VAV box looked at by a test-and-balance contractor. We found that 75 percent of them were out of calibration; some were even disconnected and others had filters that had never been replaced.

Zanolini points out some basic design choices in the original building design that lead to higher energy use, which are very difficult to fix in a retrofit.

What brings in fresh air to this building is a single fan on the roof. What exhausts the return air out of the building is also a single exhaust fan on the roof. Both of those fans run 24 × 7. If you know anything about building economy, you know that anything that runs 24 × 7 is not efficient. There was no thought of modulation.

What was the final result from Caterpillar Financial's LEED-EB project? Documented reductions in resource consumption included:

- 6 percent electricity savings
- 22 percent water savings
- 49 percent solid waste recycling level

*www.ge.com/en/company/companyinfo/quality/whatis.htm, accessed July 7, 2009.

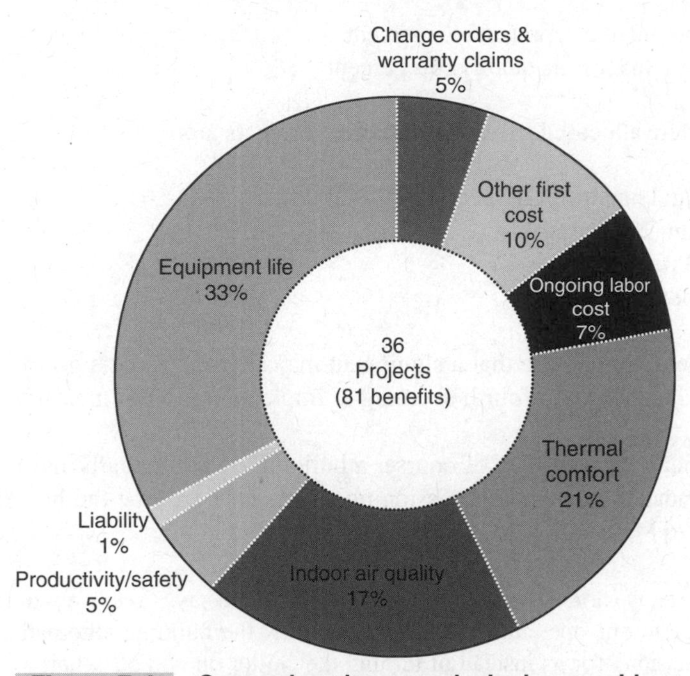

Figure 7.4 Comprehensive commissioning provides many benefits beyond just savings on energy costs, though these tend to provide the economic justification for the process.

To get these savings, few capital improvements were made. Although the original building had installed automatic faucets, urinals, and water closets, Zanolini did purchase and install low-flow, 0.5-gallon-per-minute faucet aerators, to further reduce water usage.

What are the other benefits of commissioning? Fig. 7.4 shows some of the non-energy benefits, based on a number of studies collected and analyzed by Lawrence Berkeley National Laboratory.* More than 70 percent of the benefits cited come from extended equipment life, enhanced thermal comfort for occupants, and increased indoor air quality.

According to the same study, the median energy savings from commissioning were 15 percent with average savings of 18 percent.† Survey respondents cited the following four primary reasons for existing buildings commissioning:‡

- Save energy: 94 percent
- Improve energy system performance: 69 percent

---

*Evan Mills, Norman Bourassa, Mary Ann Piette, Hannah Friedman, Tudi Haasl, Tehesia Powell and David Claridge, 2004, *The Cost-Effectiveness Of Commercial-Buildings Commissioning: A Meta-Analysis of Energy and Non-Energy Impacts in Existing Buildings and New Construction in the United States*, Report LBNL–56637 (Rev.), p. 43. http://cx.lbl.gov/cost-benefit.html, accessed May 27, 2009.

†Ibid., p. 39.

‡Ibid., p. 34.

- Ensure/improve thermal comfort: 65 percent
- Ensure adequate indoor air quality: 59 percent

Costs for Cx were allocated in existing building projects along the following lines:*

- Investigation and planning: 69 percent of total costs
- Implementation: 27 percent
- Verification: 2 percent
- Reporting: 2 percent

From the survey, one can see that a significant majority of Cx costs go to planning and fieldwork, with about one-fourth accruing to implementation of measures recommended by the Cx project.

To get the benefits of retro-Cx, of course, a building owner actually has to follow up the recommendations of the commissioning agent. That may be the hardest thing to do, according to McCown:

> The biggest barrier is for an owner to be able to trust us and say, "Yes, we will implement this change to our operations. Yes, we will allow the building automation system to make decisions for us instead of turning the chiller on and off when we think it should be turned on and off; instead, we'll allow the [computer] program to make the decision." That's one of the things that we've seen as our biggest hindrance. If they are engaged with us in a Continuous Commissioning process, that means they've said, "We want you to take us through the whole process, train us and help us realize these savings." With that we can sit down with them and help them get comfortable with not doing so much [active intervention]. It's really better to just leave things alone, allow the system to decide when to turn the chiller on and off and let the system decide what the supply air temperature should be. It's helping them get comfortable with those changes. That's the biggest barrier in implementing anything.

McCown's comments suggest that beyond technical skill, the Cx agent also needs the "bedside manner" of a skilled physician, since often the hardest part may be to convince the building manager or operator *not* to respond to every complaint that comes in about temperature or air flow, just as a good doctor doesn't prescribe drugs for every ailment, knowing that the human body tends toward health when left alone.

Economic Benefits of Retrocommissioning

What are the economic benefits of retro-Cx and what specific measures tend to be recommended? We asked Len Beyea, a certified energy manager, certified building

---

*Ibid., p. 35.

commissioning professional, and principal at RetroCom Energy Strategies to share some project information. The clients are confidential but the information represents the firm's real-world experience. Consider the example of a 25-story office building, with projected annual savings of $106,955 from an investment of $279,100. In this case, there is a 2.6-year simple payback before local utility incentives, and a 1.9-year payback after incentives. Table 7.3 shows similar data from four RetroCom projects, all with similar results. It's no wonder that retro-Cx is so viable.

In this particular office retro-Cx project, two measures requiring 87 percent of the capital cost deliver more than 80 percent of the savings, but they have higher payback periods than a variety of simpler measures. Most energy retrofits show similar characteristics: a variety of no-cost, low-cost measures yield good savings, but to get significantly improved total energy use reduction, it's necessary to invest larger sums. What Table 7.3 doesn't show is that the project found nearly $73,000 in utility incentive payments that lowered the overall payback period to 1.9 years. Every project team should be on the lookout for similar incentives, whether federal, state, or local, that will make projects more financially feasible.

TABLE 7.3 RETRO-CX FOR A 25-STORY OFFICE BUILDING*

MEASURE	PEAK kW SAVINGS	ANNUAL SAVINGS ($)	IMPLEMENTATION COST ($)	SIMPLE PAYBACK (YEARS)
1. Reset chilled water temperature based on outside wet bulb temperature	21.0	3,742	1,770	0.5
2. Optimize start time of fans and pumps	—	9,143	5,220	0.6
3. Optimize economizer cycle and CO_2 controls	13.3	2,922	1,980	0.7
4. Reset chilled water differential pressure	1.2	650	1,000	1.5
5. Add occupancy sensor lighting controls	—	53,056	126,330	2.4
6. Control open chilled water loop pumps with VFDs	1.9	4,871	12,110	2.5
7. Convert interior zones to variable air volume (VAV)	15.1	30,248	115,680	3.8
8. Control chiller pumps with VFDs	7.9	2,323	15,010	6.5
Total	**60.5 kW**	**106,955**	**279,100**	**2.6**

*Information provided by RetroCom Strategies, www.retrostrategies.com, accessed May 27, 2009.

TABLE 7.4 FOUR RETRO-CX PROJECTS, COSTS AND BENEFITS

PROJECT	NO. OF MEASURES	ANNUAL UTILITY SAVINGS ($)	IMPLEMENTATION COST ($)	SIMPLE PAYBACK (YEARS)
25-story office	8	106,955	279,100	2.6
5-story office	12	81,685	32,520	0.4
48-story office	11	103,208	135,525	1.3
850,000-square-foot hospital	23	771,929	414,802	0.5

Table 7.4 shows four of RetroCom's retro-Cx projects, with the anticipated energy savings and payback periods. In each of these projects, electric utility energy conservation incentive payments reduced payback periods even more than shown.

Technical Issues in Retro-Cx

Because of its centrality to the entire LEED-EB process and its importance in saving energy for buildings, it's important to look at some of the technical challenges in implementing retro-Cx. Here's the testimony of Darren Goody, an experienced commissioning professional with one of the leading LEED-EB consulting firms, Green Building Services:*

> Probably some of the biggest "gotchas" and findings that we come across in retro-Cx are within the control system. For example, a sequence of operations that are listed in a "controls-as-built" diagram or on a mechanical engineer's drawing do not begin to really detail out exactly how the direct digital controllers are programmed. They provide just a hint of the actual sequence. Once you start digging into how an air-handling unit functions or how a chiller plant stages itself up and down, you begin to uncover a lot more detail than what was present in the written sequence.
>
> Every time we run through a retrocommissioning effort, we develop a much better understanding of how the controls are installed and programmed. We come across things that don't quite make sense based on how they're programmed. Easy tweaks either in the code or in set points can be modified to save energy. For example, a building has what's called the morning cool-down, morning purge or night high limit. The premise is that during the coolest part of the morning, usually right before the building goes into occupied mode, outside air is utilized for an air-side economizer-equipped air handling unit to cool the space and purge the internal gains left over from

*Interview with Darren Goody, April 2009.

> the previous day (because the building envelope continues to radiate heat through the night). Instead of ramping up the chiller plant right away in the morning, that outside air is used to precool the building and thereby save chiller plant energy. The more mass a building contains, the more it makes sense to utilize a morning purge cycle. But even when it's already programmed to do so, there are tweaks that can be made to the set points to make it more efficient and make it operate more frequently. That's just one example; whatever system you're talking about, we can find control improvements in the retrocommissioning process.

Goody's statement illustrates that saving energy in actual building operations is as much of an art as a science. An experienced engineer can always find something that either should have been designed better in the original building or that is no longer working as designed. In addition, Cx agents often find things that are broken and have never been fixed, such as air dampers that are stuck open or closed.

Of course, even retro-Cx can't deliver miracles in a well-run building. The Sacramento Municipal Utility District (SMUD) is a public agency in California that applied a LEED-EB process to their Customer Service Center in Sacramento. Doug Norwood, a senior mechanical engineer, says:*

> The building's energy use was already about 30 to 35 percent below California's Title 24 [Energy Code] before we started the LEED-EB project. As a result of the project—accounting for only the very specific things that we did just for LEED—we were probably able to pull that [use] down another 2 to 3 percent. It was a very efficient building already, and over the years we've found that it's difficult to squeeze more savings out of it. We did a retrocommissioning [with outside consultants]; they had some good recommendations for us, so we implemented just about all of those and saw some savings come out of that.

Energy-Efficiency Measures for Existing Buildings

Obviously the purpose of commissioning is to find out what to do to upgrade the energy performance of a building. Many analyses focus on "site energy," that is, energy directly consumed on-site. However, it is important to consider energy upgrades from the broadest perspective—the generation of electricity for buildings at power plants. About 67 percent of all U.S. electricity production goes to all types of buildings.† The energy used to produce electricity for a particular building is known as *primary energy*. About 60 percent of the primary energy is lost in fossil-fuel electric generation, and an additional 10 percent is lost in transmission between the plant and the building, leaving the

*Interview with Doug Norwood, March 2009.

†www.eia.doe.gov/cneaf/electricity/epa/epa_sum.html, Table 7.2, accessed July 9, 2009.

TABLE 7.5 SOURCE ENERGY TO SITE ENERGY RATIO, U.S. AVERAGES*

FUEL TYPE	EPA PORTFOLIO MANAGER DEFAULT
Electricity	3.34
Natural gas	1.05
Steam	1.45
Hot water	1.35
Chilled water	1.05

*LEED Reference Guide, p. 162

overall system at about 30 percent efficiency at best.* Table 7.5 shows the primary (source) to end-use (site) energy ratio, compared to end-use energy consumed, for the most-used energy sources in buildings. This table shows that electricity consumes 2 to 3 times the primary energy of other fuel sources and other end-uses.

The Energy-Efficiency Market

Spending on energy-efficiency upgrades in the United States in 2004 was estimated at $300 billion, with a "premium" for efficiency upgrades (beyond what would normally be spent) estimated at $43 billion (about 15 percent of the total expenditure.)† This estimate includes spending by buildings, industry, transportation, and utilities. Looking at buildings alone, the market size was estimated at $178 billion (about 60 percent of the total efficiency market), with $24 billion representing an energy-efficiency premium. The annual value of building energy savings was estimated at $12 billion, representing a two-year payback at today's prices on the efficiency premium of $24 billion.‡ For commercial buildings, the estimate is $51 billion in total investment, with about a $7.7 billion premium.

In 2004, for example, investments in efficient appliances and electronics represented nearly half the total (48 percent) of all building-related investments, vastly

*http://climatetechnology.gov/library/2003/tech-options/tech-options-1-3-2.pdf, accessed May 29, 2009.

†Karen Ehrhardt-Martinez and John A. "Skip" Laitner, "Size of the U.S. Energy Efficiency Market: Generating a More Complete Picture," American Council for an Energy-Efficient Economy, Report Number E083, May 2008.

‡"Payback period" is a common term that denotes how long it takes to recover an initial investment in annual or monthly savings. Payback doesn't deal with the long-term economic return of an investment, but only specifies when the initial capital expenditure is recovered. Most American businesses have a payback criterion that typically limits discretionary investments to paybacks of 18 to 30 months, ruling out many good energy-efficiency retrofits with longer payback periods. So it is best to use additional terms to analyze and characterize these expenditures, such as return on investment, net present value, or even "five-year return" to sell energy-efficiency solutions.

TABLE 7.6 ECONOMICS OF ENERGY-EFFICIENCY AND NEW ENERGY RESOURCES*

NEW ENERGY RESOURCES	LEVELIZED COST (CENTS PER KILOWATTHOUR)
1. Energy efficiency	3.2
2. Wind	6.6
3. Biomass	7.2
4. Natural gas combined cycle	8.5
5. Pulverized coal	10.4
6. Thin-film photovoltaics	10.8
7. Nuclear	11.1
8. Solar thermal (concentrating)	11.8

*American Council for an Energy Efficiency Economy, "Energizing Virginia: Efficiency First," September 2008, Figure ES-1, www.aceee.org/pubs/E085.htm, accessed September 19, 2008.

exceeding their estimated energy use in buildings (about 8 percent.) In other words, people will make the "easiest" investments first, buying products and technologies not requiring significant building upgrades, such as buying ENERGY STAR computer monitors and ENERGY STAR printers and copiers.

Key factors driving all energy-efficiency investments include:

1. The positive economics of energy efficiency as a reliable new source of energy, compared with all available alternatives, as shown in Table 7.6.
2. Higher costs of oil and more volatile prices for gasoline and natural gas (electricity prices do not move quickly, typically owing to the state regulatory apparatus controlling them).
3. Concern that energy supplies may be interrupted or curtailed by rising demand charges (time-of-day charges are based on a building's or facility's peak 15-minute power demand in a month or quarter, especially during periods of peak demand on utility resources) or by "coerced" conversion to interruptible supplies (local electric utilities may place controls on electric service in a place of business that shut off service in the case of grid difficulties).
4. Policy changes by businesses and large institutions to respond to the urgency many feel about the global warming challenge (investing in energy efficiency may be done as a priority investment, compared with other uses of capital, as evidenced by more than 900 U.S. mayors and more than 600 college and university presidents committing to dramatically reduce greenhouse gas emissions).*

*See, for example, the U.S. Conference of Mayors, www.usmayors.org, and the Association for the Advancement of Sustainability in Higher Education, www.aashe.org.

5 Growing consumer and stakeholder pressures, particularly on public companies and public institutions, to invest in energy conservation and efficiency.
6 New technologies, products, and systems that are making efficiency investments more cost-effective.
7 Increased availability of incentives and financing mechanisms for energy-efficiency and renewable energy upgrades.*

INCENTIVE PROGRAMS

In 2008, the American Council for an Energy-Efficiency Economy released its state energy-efficiency program "top 10" list. The list documents best practices and leadership among states focused on energy efficiency. The states are ranked on their adoption and implementation of energy-efficiency policies and practices. The top 10 states with incentives in 2008 were:†

1 California
2 Oregon
3 Connecticut
4 Vermont
5 New York
6 Washington
7 & 8 Massachusetts and Minnesota (tie)
9 Wisconsin
10 New Jersey

LIGHTING EFFICIENCY RETROFITS FOR EXISTING BUILDINGS

The federal Energy Policy Act of 2005 (EPACT) provides a federal tax deduction of $0.30 to $0.60 per square foot for lighting retrofits in commercial buildings, that result in a 50 percent reduction in energy use, as measured by the ASHRAE 90.1-2001 standard. The law also provides up to $0.60 per square foot each for HVAC upgrades and building envelope upgrades (such as replacing windows).‡

In 2008, the EPACT law was extended through 2013. This allows plenty of time for building owners to design and install energy-retrofit projects. Regardless of federal tax benefits, the economics of energy-efficiency upgrades are compelling in most areas of the United States, with paybacks of three years or less common for lighting efficiency upgrades. This owes in part to electric utilities (most investor-owned and some municipals) that offer significant financial incentives for energy-efficiency upgrades.

*See, for example, the Database of State Incentives for Renewable Energy and Conservation, www.dsireusa.org.
†Maggie Eldridge, Max Neubauer, Dan York, Shruti Vaidyanathan, Anna Chittum, and Steve Nadel. "The 2008 State Energy Efficiency Scorecard," American Council for an Energy-Efficient Economy, October 2008.
‡Database of State Incentives for Renewable Energy and Energy Efficiency, http://www.dsireusa.org, accessed July 7, 2009.

An additional legislative market push came from the Energy Independence and Security Act (EISA) of 2007. Section 324 of the law set a new federal standard for efficacy of the types of ballasts used in metal halide lamp fixtures. Also, for the first time Section 321 set an efficiency standard for "general service" light bulbs to phase-out the most common types of incandescent light bulbs by 2012 to 2014. The regulation is not a product "ban," but rather a performance requirement for power consumption, light output, and lifetime. This law should accelerate the development of new lamp technologies to meet the increasingly stringent federal standards. The EISA law also put the onus on the federal government to reduce energy use in its buildings 3 percent per year from 2008 to 2015.*

The American Recovery & Reinvestment Act (ARRA) provided approximately $75 billion in potential funding for energy efficiency, renewable energy, and the smart grid.† About $26 billion will focus on energy-efficiency–related improvements, mostly in buildings.‡ This record amount of funding and tax incentives will be the single largest determinant of energy management growth over the next several years. Many of these funds will be distributed over the next two to three years because the overall goal is to stimulate the economy in the near-term. The funds will be channeled through federal and state agencies. Energy-efficiency–related spending will occur through both direct and indirect funding mechanisms.

To summarize, some of the factors driving energy efficiency market adoption include rising electricity prices (especially for peak periods), a growing number of utility and state efficiency incentives, increasingly stringent energy code requirements for new buildings and a growing awareness among building operators about the need to cut operating costs for energy.

Importance of Controlling Lighting Energy Use

Lighting represents roughly 25 to 40 percent of the total commercial building energy consumption in the United States§ (vs. 10 to 15 percent in residential),¶ and therefore represents a natural place to look for significant efficiency gains. In 2003, lighting

---

*www.nema.org/gov/energy/upload/NEMA-Summary-and-Analysis-of-the-Energy-Independence-and-Security-Act-of-2007.pdf, accessed September 21, 2008.

†The entire ARRA encompasses $787 billion in economic recovery spending programs, www.recovery.gov, accessed May 26, 2009.

‡Kateri Callahan, "Energy Efficiency & the Stimulus Bill: Rebuilding the Economy Today for a "Green Energy" Tomorrow." Alliance to Save Energy webinar, March 17, 2009.

§Marc Hoffman and Ed Wisniewski, "Business Opportunities in the Commercial Lighting Arena: High-performance T8s light the way for savings opportunities on new and existing construction projects," Consortium for Energy Efficiency, April 1, 2008.

¶Jerry Yudelson, *Choosing Green: The Homebuyers Guide to Good Green Homes*, Gabriola Island, BC: New Society Publishers, 2008.

comprised almost exactly 25 percent of delivered (end-use) energy in office buildings (space heating, cooling, and ventilation made up just about 50 percent of the total), while it was 28 percent (enclosed malls) to 35 percent (free-standing) in the retail sector.*

However, a focus on lighting energy retrofits alone will not yield as many benefits as a more comprehensive look at the entire building or facility. For example, evaluations of first-generation *comprehensive* retrofit programs report whole building energy savings of 11 to 26 percent of pre-retrofit consumption compared to 8 to 13 percent savings for comprehensive lighting retrofits that did not involve other end uses.†

Another analysis of 678 projects completed by energy service companies (ESCOs) between 1982 and 2000 found median project costs of $2.50 per square foot for institutional projects and $1.40 per square foot for private sector projects. As a whole, institutional projects installed a greater number of efficiency measures compared with commercial projects, accounting for the higher costs. Overall, in 94 projects incorporating lighting and non-lighting measures, median electricity savings were 23 percent of total facility electricity use; in 63 lighting-only projects, median savings totaled 47 percent of lighting electricity use, representing about 12 to 14 percent of total facility electricity use.‡

RETROFIT OPPORTUNITIES

By one estimate, obsolete T12 lamps§ still account for more than 50 percent of the total number of fluorescent lamps in use today in the United States.¶ According to one study, there are about 1.5 billion fluorescent lamps installed in the commercial sector.**

Of lighting consumed by fluorescent fixtures, T12s consume about 70 percent of the total. Converting all T12s to high-performance T8s could save about one-third of this energy use.†† (This estimate speaks only to lamps and does not consider how dimmable ballasts and occupancy sensors can further add to the total savings from lighting energy retrofits.)

In many retrofit situations, reduced-wattage systems might be better suited where fixture placements will not change. High-performance T8 lighting systems are more efficacious, but reduced wattage lamp-ballast systems still offer considerable potential for energy savings where facility managers are unlikely to use high-performance systems.

---

*http://buildingsdatabook.eren.doe.gov/, accessed July 7, 2009.

†Jennifer Thorne Amann and Eric Mendelsohn, "Comprehensive Commercial Retrofit Programs: A Review Of Activity And Opportunities," April 2005, Report Number A052, American Council for an Energy-Efficient Economy.

‡Ibid.

§The fluorescent lamp that is most often used for both general ambient and specific task lighting is the "energy saver" four-foot, cool white, fluorescent tube (F40T12/CW/SS). http://ateam.lbl.gov/Design-Guide/DGHtm/historyandproblemsoft12fluorescentlamps.htm, accessed July 7, 2009.

¶A T12 lamp is 12/8 inches in diameter, or 1.5". A T8 lamp is 8/8 inches in diameter, or 1 inch, and a T5 lamp is 5/8 inches in diameter.

**Marc Hoffman and Ed Wisniewski, "Business Opportunities in the Commercial Lighting Arena: High-performance T8s light the way for savings opportunities on new and existing construction projects," *op. cit.*

††Ibid.

Incentive programs for lighting retrofits Because they are so easy to retrofit, putting in lower-energy-using linear fluorescent lamps such as high-efficiency T8s or installing compact fluorescent lamps (CFL) lamps for incandescent fixtures (such as recessed "cans" in ceilings) will save both energy and maintenance labor costs in most settings. As a result, many electric utilities offer strong incentive payments for commercial users to carry out these retrofits.

Al Skodowski is senior vice president and director of LEED and sustainability for Transwestern, a national commercial real estate services and development firm currently pursuing LEED certification for more than 15.4 million square feet of new and existing commercial space. In his experience, lighting retrofits are a "no brainer," even if there are no utility incentives available.*

> Another thing we are routinely finding is that new lamp technologies have lower mercury content with longer run hours, up to 40,000 hours at 28 watts, whereas a typical T8 lamp is 32 watts, 20,000 hours. You can effectively put those in to earn the [LEED-EB] low mercury points and save four watts a light bulb. One building in Coral Gables, Florida, that Transwestern manages has 1100 lamps so when you multiply that times four watts, it is a pretty good savings (4.4-kW). In some cases, we have been able to move them down to 25-watt T8 lamps. Because Transwestern is working on these buildings, we are able to apply a fair amount of education to the process for both property managers and chief engineers. As a result, we save them operating expense dollars. When you look at a 20,000-hour bulb versus a 40,000-hour bulb, guess which one you are changing more often. That incurs maintenance dollars on top of it.

Emerging Energy Retrofit Technologies

Lighting controls and sensors, building controls, variable-speed drives, LEDs (light-emitting diodes), and solar power installations represent emerging technologies for the energy retrofit market.

LIGHTING CONTROLS AND SENSORS

In a survey of electrical manufacturers and distributors conducted by the author, 44 percent of manufacturers said lighting control systems are the main product family for which they are seeing increasing demand.† Occupancy sensors and dimmable ballasts also showed almost the same level of demand.

Researchers at the California Lighting Technology Center (CLTC) tested the utility and efficiency of integrated lighting and control systems. They investigated the point

*Interview with Al Skodowski, March 2009.

†Yudelson Associates, "Green Goes Mainstream: How to Profit from Emerging Market Opportunities," for the National Association of Electrical Distributors, Oct. 2008, www.naed.org/Education%20and%20Research%20Foundation/Foundation%20Research/Foundation%20Research%20Completed%20Research.html, accessed July 7, 2009.

at which efficiency jeopardizes fundamental lighting quality. The integrated office lighting system tested by the CLTC combined low-level ambient lighting and under-cabinet LED task lighting controlled by an occupancy sensor. The CLTC demonstrated this system can achieve a 0.5 to 0.7 watt/square foot lighting-power density, which is 36 to 55 percent lower than ASHRAE 90.1-2007, the current LEED-NC 2009 baseline standard.*

In another high-profile example, the lighting control system at the New York Times' building in New York City is outstripping its forecasted 70 percent reduction in energy use, by delivering savings that amount to $30,000 per year per floor (about $1.00 per square foot), according to performance figures released in May 2009.† The New York Times' 52-story tower, with 1.6 million square feet of Class A office space, opened in late 2007. Initial information gathered on the lighting system since the opening showed energy savings of 72 percent on the floors where it had been installed, according to Glenn D. Hughes, the director of construction for The New York Times Company. The cost savings resulted from a significant reduction in both direct lighting use and indirect cooling load.‡

While the New York Times example represents a new building, the use of comprehensive building lighting controls should offer significant savings for older buildings. One caveat, however, by reducing lighting loads significantly, an older building's HVAC system may not run enough (because far less cooling is needed) to provide code-required ventilation levels.§ Again, this example shows why a comprehensive retro-Cx program is needed, along with whole building simulation modeling for larger buildings, as illustrated by the Empire State Building example in Chap. 1.

Occupancy sensors Occupancy sensors use infrared or ultrasonic technology to detect activity. Infrared sensors generally do not "see" as far as ultrasonic units. Dual-technology sensors cost more but combine both methods for increased accuracy and flexibility. Energy savings from occupancy sensors fall within the following ranges for various applications:¶

- 40 to 46 percent in classrooms
- 3 to 50 percent in private offices
- 30 to 90 percent in restrooms
- 22 to 65 percent in conference rooms
- 30 to 80 percent in corridors
- 45 to 80 percent in storage areas

*Craig DiLouie, "Are Highly Integrated Lighting and Control Systems the Future of Lighting?" *Archi-Tech Magazine*, July/August 2008.

†www.lutron.com/news/press_release.aspx?cid=0&ArticleId=417, accessed May 28, 2009.

‡www.greenerbuildings.com/news/2009/05/28/high-efficiency-lighting-new-york-times-building-exceeds-expectations, accessed May 28, 2009.

§Mark Heizer, "Saving Energy in Office Buildings," *Heating Piping and Air Conditioning Magazine*, May 2003

¶http://ecmweb.com/lighting/electric_occupancy_sensors/, accessed June 24, 2008.

Dimmable ballasts There is a potential nuisance factor for electric lighting controlled by photosensors–sudden changes in lighting levels. As a result, most daylighting integration approaches use dimmable ballasts. These controls need to be commissioned to ensure they meet design intent and occupant acceptance.* Most lamps, including incandescents, when dimmed by 25 percent will use approximately 20 percent less energy. Light dimming also reduces thermal load on a building's cooling system, which increases energy savings.† Dimmable ballasts vary the light output of fluorescent lamps to lower the light level whenever full illumination is not needed. However, fluorescent lamp dimming controls require more wiring than incandescent lamp dimmers unless wireless technology is used.

DAYLIGHTING INTEGRATION

More than 100 years ago, Albert Einstein explained the photoelectric effect,‡ now the main source of design for photosensors and lighting controls. Controlling lighting in buildings in conjunction with daylight means using some form of light sensor.

Lighting represents a major energy end use in commercial buildings (typically about one-third of the total) and also affects cooling and heating loads. Heat generated by lighting, equipment, and occupants in a commercial or institutional building often produces a cooling load for most of the year during daytime occupancy hours.

One method frequently employed to reduce lighting energy demand is to increase the use of daylight and also to use daylight-responsive lighting controls, provided solar heat gain and glare are also controlled.

Using higher-efficiency lamps and ballasts, improving the effectiveness of fixtures, and creating a coordinated space layout collectively increase the efficiency of building lighting and reduce carbon dioxide emissions from electric power generation. Using daylighting systems with appropriate shading and electric lighting controls adds to those energy and cost savings by reducing lighting energy consumption and moderating peak demand in commercial buildings.

To achieve best results, retrofits need to zone interior spaces for optimal placement of luminaires and sensors, with luminaires typically placed parallel to the windows and circuited with respect to distance from daylight sources, including skylights. Lighting needs to be referenced specifically to the location of workspaces. In this respect, the lighting designer should consider both task and ambient lighting when deciding on the best efficiency upgrades.

By using controlled daylighting and other effective lighting controls, the energy use of lighting can be reduced below 1.0 watt/square foot, even as low as

*DiLouie, *op. cit.*

†www.lrc.rpi.edu/programs/lightingTransformation/residentialLighting/buildersGuide/controls.asp, accessed September 1, 2008.

‡www.einsteinyear.org/facts/photoelectric_effect/, accessed May 27, 2009.

0.5 watts/square foot, which represents the level of a typical design in a green or high-performance building. Lighting controls also reduce the air-conditioning load on a building and capital costs for green buildings. Finally, lighting controls reduce the incidental and unintended heating of the building resulting from inefficient lighting design.

To receive credit for daylighting in LEED-EBOM, 50 percent or more of regularly occupied spaces must exhibit daylight illumination levels of at least 25 foot-candles (270 lux) (and no more than 500 foot-candles).* Projects also must incorporate glare control devices to avoid high-contrast situations.

An architect friend of the author's often makes the case for investing in daylighting to skeptical school officials by taking them on tours of daylit schools so they can "feel" the difference, and by pointing out the operating costs are lower and lamps are replaced less often, saving on labor costs. He strongly believes that daylighting is also healthier for teachers, staff, and students.†

In the commercial market, research shows daylight increases retail sales by about 5 percent, a good reason for building retrofits and renovations to consider installing skylights. A study conducted by the California Energy Commission used statistical analysis to show a positive correlation between daylighting and retail sales. Higher illumination levels, improved color rendering, greater depth perception, and biological effects are some of the reasons why daylighting improves sales results.‡

In stores with very high ceilings, daylighting controls can be simple. In one such case known to the author, an engineer specifying equipment for large new stores of a grocery chain designed three-lamp fluorescent fixtures. Oriented in specific zones under skylights, the fixtures were designed with multiple ballasts in series to operate with one, two, or three lamps turned on (or, on a really bright day, all turned off). Since the lamps were high up near the ceiling and the shoppers were moving about, the abrupt change from one lamp on to two lamps on did not matter to shoppers. Additionally, by comparing lamps in one fixture to nearby fixtures on the same zone, maintenance people could see if a lamp had burned out and quickly replace it.§

Table 7.7 shows the number of different types of lighting control strategies often used in commercial and institutional buildings. Daylighting alone offers noteworthy energy savings, but when daylighting is integrated with controls and sensors, then the potential for savings increases significantly.

Daylighting controls can be retrofitted into existing buildings, along with other forms of lighting controls. Such controls may reduce electricity use by 30 to 40 percent

---

*LEED Reference Guide, pp. 409ff.

†Capital E Consultants, www.cap-e.com/ewebeditpro/items/O59F3303.ppt#2, accessed March 6, 2007.

‡Lisa Heschong, "Daylight and Retail Sales," prepared by Heschong Mahone Group, Inc. on behalf of the California Energy Commission, October 2003.

§Personal experience of the author.

TABLE 7.7 LIGHTING CONTROLS AND ENERGY SAVINGS POTENTIAL*

LIGHTING CONTROL METHOD	WHAT IT DOES	ENERGY SAVINGS POTENTIAL
Daylighting controls	Provides constant lighting levels by reducing electric lighting as daylight increases	35%, in areas with good daylighting (replaces electric lighting)
Task lighting	Allows occupant control of lighting levels	5%
Vacancy or occupancy sensors	Turns lights off when vacant; on when occupants detected	30% to 45% in zoned spaces, such as private offices
Scheduling	Automatic shutoff according to prescribed occupancy schedules	25%
Integrated strategies	Combines all measures	50% or more

*http://www.automatedbuildings.com/news/sep07/articles/schwartz/070828022606schwartz.htm, accessed September 29, 2008.

in the row of lights closest to the windows and 16 to 22 percent for the second row of lights. Paybacks can be as little as three years on such investments.*

Lighting controls also save on building operating costs. For example, occupancy sensors combine ultrasonic and infrared sensors at costs of $30 to $100 each, with payback periods ranging from six months to five years, depending on initial cost, occupancy behavior, and local electricity cost.† Some studies show workers are out of their offices 30 to 70 percent of the time during working hours. A conservative estimate of savings possible from controls is about 30 percent.‡

Case Study: Chicago Transit Authority (CTA) Headquarters, Chicago

Managed by Transwestern, the 400,000-square-foot CTA Headquarters building went through a LEED-EB certification effort in 2006 (Fig. 7.5), the first building in Illinois to receive the award.§ In early 2007, the project was certified at the Gold level. As the

*See chapter on daylight responsive controls from Lawrence Berkeley National Lab, http://gaia.lbl.gov/iea21/documents/sourcebook/hires/daylighting-c5.pdf, p. 5, accessed September 14, 2008.

†North Carolina Energy Office, www.p2pays.org/ref/32/31316.pdf, accessed September 14, 2008.

‡http://gaia.lbl.gov/iea21/documents/sourcebook/hires/daylighting-c5.pdf, p. 7, accessed July 18, 2008.

§www.transwestern.net/fact_sheet/sustainability.pdf, accessed May 28, 2009.

Figure 7.5 Constructed in 2006, the CTA Headquarters building went through the LEED-EB process in 2006 and raised its ENERGY STAR score from 83 to 87. *Photo by John Herbst, courtesy of Transwestern.*

building was constructed in 2004, most of its systems were still new, and the building had an ENERGY STAR score of 83 already. Transwestern implemented the following efficiency strategies:

- Fine-tuned all mechanical systems through the commissioning process, resulting in a higher ENERGY STAR score in 2006
- Upgraded plumbing fixtures, which reduced water consumption by 33 percent from the LEED baseline water consumption
- Implemented a full recycling program
- Installed a 29,000-square-foot green roof, to reduce the Urban Heat Island Effect and to capture stormwater for later reuse

- Provided greater daylighting through the use of lighting controls, so that 75 percent of all occupants have access to natural light
- Maintained views to the outdoors from more than 80 percent of all regularly occupied spaces
- Created storage and shower facilities for bicycle commuters
- Separately ventilated janitor closets keep odors and emissions from cleaning chemicals out of the building air
- Instituted specific low-impact green cleaning programs, along with integrated pest management, to reduce occupant exposure to chemicals

The project showed quantifiable results that reduced costs for the CTA, including an increase in the ENERGY STAR score from 83 to 87 in 2006, a reduction in overall operating expenses by 1.4 percent (because of a 12 percent reduction in utility costs), and reuse of about 500,000 gallons of rainwater annually captured and slowly released by the green roof. The total cost of the project was less than $40,000. Annual upkeep and preventive maintenance programs had a total cost of less than $5000.

The CTA project was Transwestern's first LEED-EB project. After completing a number of subsequent certification projects, senior vice president Al Skodowski doesn't think LEED-EB has to be costly:*

> What we have seen from the early beginnings with the CTA [project] through current projects is that it really does not cost a lot of money. Payback periods are pretty reasonable. Generally we are seeing paybacks in the two-to-three-year range, at the top end, with operation expense reductions specifically tied to energy and then water savings. Most of the pieces are cost-neutral: green cleaning, green cleaning supplies, green cleaning equipment, pest control, exterior hardscape maintenance, etc.

BUILDING CONTROLS

About 28 percent of electrical manufacturer respondents to a survey conducted by Yudelson Associates in 2009 see increasing demand for building controls and building automation systems (BAS). In the same survey, 60 percent of electrical distributors said variable speed drives and/or motors represent the product family with the greatest increase in demand. An intelligent building contains an infrastructure allowing information to be exchanged between various building systems. BAS is a control system based on a computerized network of electronic devices designed to monitor and control the mechanical and lighting systems. Communications and Internet connectivity are critical to all BAS.†

In addition to the HVAC system, various other types of equipment can be connected to a BAS, including video-conferencing equipment, video projectors, lighting control

*Interview with Al Skodowski, March 2009.

†Karen Kroll, "A Measure of Building Intelligence," *Building Operating Management*, September 2007.

systems, public-address systems, and touch-screen kiosks with displays of energy and water use.*

Wireless controls and sensors are playing a growing role in building automation systems. Overall demand for wireless technology is increasing as well. Wireless systems are especially well suited to retrofits as they are easier to install than hardwired controls and sensors. Applications include temperature sensors for ongoing commissioning and occupant comfort, occupancy sensors for lighting, energy metering, and load control.[†]

Integrated intelligent building systems reduce operating expenses and give facility managers a unique method of improving a building's operating efficiency.[‡] Building control software gives facility managers a concise picture of what systems use the most energy, allowing them to identify energy saving opportunities. This software also allows them to coordinate the lighting, HVAC, and access control systems so that these systems can be powered down or turned off when the building is unoccupied. An integrated automation system also increases a building's value, attracting higher rents and quality tenants.[§]

Using a Web-based platform for demand response programs allows a building owner to turn off nonessential equipment (unused computers, printers, monitors, etc.) during peak demand periods, reducing utility charges and the strain on the grid and the local distribution system.

Sears Tower in Chicago, the largest office building in the United States at 1450-feet tall (and the third largest in the world) is participating in a demand response program from a private vendor, EnergyConnect. The platform, called FlexConnect, works by upgrading the building automation system to provide more timely data for demand response, allowing participants to control their energy costs and receive incentive payments by voluntarily reducing electricity use in response to high market pricing or regional shortages.[¶]

Adding building automation systems (BAS) to a major renovation requires a large initial investment. However, these systems save significant operating expenses over the future life of the building. With BAS systems, HVAC and lighting equipment will last longer, require less maintenance, and use less energy, resulting in substantial savings.** In addition, it is impossible to manage large buildings with complex systems without a modern BAS in place.

Mark Frankel, technical director at the New Buildings Institute, thinks that a BAS can be used more effectively to help building managers and tenants reduce their energy consumption:[††]

---

*www.facilitiesnet.com/bas/, accessed June 13, 2008.

[†]Interview with Bob Heile, July 28, 2008.

[‡]Interview with John Jennings, market manager with the Northwest Energy Efficiency Alliance, Portland, Oregon, August 2008.

[§]Kroll, *op. cit.*

[¶]www.bdcnetwork.com/articleXml/LN850408241.html, accessed September 20, 2008.

**Ibid.

[††]Mark Frankel interview, January 2009.

Simpler controls are often adequate, even in complicated buildings. What you need is functional feedback about the building's performance. You basically need to measure temperature, occupancy and base loads, lighting run times, and total energy use. It just takes these four measurements to compare consumption across different buildings. These four measurements are indicators that determine whether you need to go deeper to find inefficiencies. *They are the key performance indicators*. What you want building controls and energy management systems to produce is actionable information for the designers, operators, and tenants. This doesn't have to be real-time (streaming) data, but it should be more frequent than once a month or once a year.

In general, the building operators either see too little or too much information. What they need is to be able to compare the performance of their building to other buildings. Building operators need sensible, transparent solutions. If tenants don't see data that is separated out for their part of the usage, they have no way of seeing the impact of their behavior. When they are comparing lease deals, they aren't presented with building performance data.

Smart meters Smart meters are digital meters that communicate energy use information between utilities and consumers. They allow utilities and energy users to access energy use information in real time. Understanding how energy is used can help utilities and consumers to save energy and money. Smart meters will be an integral component of the future smart grid.

Developing technology in this arena will allow wireless controls and sensors to harvest ambient energy sources such as radio frequency, solar, thermal, and vibration, eliminating the inconvenience of replacing batteries in wireless devices.* The U.S. Departments of Energy and Commerce are creating standards for some of these new technologies.† For example, San Diego Gas & Electric plans to have more than 200,000 Itron smart meters installed in 2009 and have its entire territory of 1.4 million customers covered by the end of 2011.‡

The ZigBee Alliance, an association of companies working to create an open global standard for these products,§ certifies a range of smart meters for residential, commercial, and industrial applications. While ZigBee meters are wireless, other companies offer smart meters to commercial and industrial markets that communicate over existing power lines.¶

Plug load controls Another way to increase energy efficiency is to put office equipment on switches that can be turned off at night to reduce plug loads.** Phantom

---

*Kurt Roth and James Brodrick, "Energy Harvesting for Wireless Sensors," *ASHRAE Journal*, May 2008.

†www.environmentalleader.com/2009/05/22/doe-doc-release-first-set-of-smart-grid-standards/, accessed May 29, 2009.

‡http://public.sempra.com/newsreleases/viewpr.cfm?PR_ID=2368&Co_Short_Nm=SDGE, accessed July 30, 2009.

§Heile, *op. cit.*

¶Mitchell Stein, personal interview, January 12, 2009.

**The term "plug load" represents electricity used by any device that gets plugged into electrical outlets. Standby power is the electricity devices consume to keep internal clocks and displays running, even when not on.

or "vampire" energy is the electricity used by a plugged-in device, even when it is turned off.* For instance, when cell phone chargers are plugged into the wall they still draw power even if they aren't charging a cell phone battery. Plug loads account for 9 percent of residential energy use[†] and 20 percent of commercial building energy use in the United States.[‡]

VARIABLE-SPEED DRIVES/MOTORS

The single largest end-use of electric power is the industrial motor. In fact, 65 percent of industrial power is used to run motors, which typically run at full speed even if they don't need to.[§] Variable speed drives (VSDs), also known as variable frequency drives (VFDs), can vary the shaft speed to match the driven load. This allows motors to speed up or slow down and to appropriately meet the power requirements at any given time. VSDs also allow motors to soft start, reducing mechanical stress on the motor and the equipment driven by the motor.[¶]

VSDs are particularly effective at improving HVAC system efficiency because they can meet the changing system needs of pumps and fans. VSDs are also used for elevators, boiler fans, and cooling towers. For a typical circulating pump, a VSD can reduce power consumption by as much as 70 percent.**

LED LIGHTS

Legislative mandates, the sustainability movement, and performance-based energy codes are driving the transition to more efficient lighting products. While there is a rapid adoption of CFLs by businesses and consumers, they are considered transitional by some in the lighting industry, many regarding LEDs as the natural next step in the evolution of efficient lighting technology. For example, LEDs are 2 to 3 times more efficient than incandescent light bulbs[††] and often double the efficacy (measured in lumens per watt) of CFLs, depending on the application.

Considering the manifold advantages LEDs offer, it is easy to understand why demand for them is increasing and why they are likely to find use in more building retrofits. The clearest advantages they offer are in high efficiency, lifespan, versatility, and easy integration with lighting control systems.

*When devices are turned off but still plugged in, they can draw power through two types of standby modes: passive standby or active standby. An example of passive standby is the clock on a microwave oven. An example of active standby is a DVD display when it is programmed to record something.

†http://www.efficientproducts.org/product.php?productID=11 accessed July 7, 2009.

‡Carol Sabo, "Plug-Load Energy-Efficiency on Campus," a presentation given at the Advanced Design and Technologies for Higher Education Facilities Workshop in Lansing, MI, September 23, 2005. PA Government Services, Inc.

§http://www.nema.org/prod/technologies/upload/TDEnergyEff.pdf, accessed March 26, 2009.

¶http://www.energy.wsu.edu/documents/engineering/motors/MotorDrvs.pdf, accessed March 26, 2009.

**http://www.energy.wsu.edu/documents/engineering/motors/MotorDrvs.pdf, accessed March 26, 2009.

††http://www.ledcity.org/about-leds/, accessed August 29, 2008.

In addition to longer life span, LEDs are highly versatile, shock and vibration resistant, mercury-free (unlike CFLs), and work well in cold environments. Their color variety lends them to "architainment" applications, where lighting and architecture are combined to attract people and hold their attention.* This was evident at the 2008 Olympic Games in Beijing, where LEDs were used on many buildings and in the opening ceremonies.

Because LEDs are computer chips, they integrate easily with lighting control systems. Lighting controls improve energy-efficiency on their own, but when paired with LEDs and daylighting, the efficiency level significantly increases. Task lighting† is an especially effective use for LEDs, as long as they are carefully integrated. (Efficiency gains could be lost if LED task lights are added to a lighting system already providing sufficient illumination.) Fixtures need to be developed specifically for LED use.‡ Because LEDs utilize a different light source than fluorescents, they work best in fixtures specifically designed for their unique attributes.

Currently, LEDs have some significant drawbacks: price, color quality, reliable manufacturing sources, and reliability. At this time, high price is the obstacle most frequently cited.§ Color quality is another obstacle: although LEDs are available in various colors, good color rendering and color temperature for white light are often lacking. In addition, LEDs are relatively unproven and do not yet have a reliable track record (meaning that efficacy in the field matches performance in the lab.)¶

Solar and Renewable Energy

Both the George W. Bush and Obama Administrations significantly incentivized renewable energy. Congress extended the investment and production tax credits (ITC and PTC), in 2005 and 2006, and approved the Emergency Economic Stabilization Act (EESA) in late 2008. In the EESA, Congress gave the PTC another three-year extension and made the ITC accessible to all forms of renewable energy under the 2009 ARRA law mentioned above.** ARRA also provides $16.8 billion in direct spending for renewable energy and energy efficiency programs during the next 10 years.††

SOLAR POWER

Renewable energy, especially solar photovoltaic systems, are potential additions to any green building retrofit. After all, what is a green building without green power?

---

*http://www.csemag.com/article/179113-Lighting_Return_of_the_Engineer.php, accessed July 30, 2009.

†Individuals at their workplaces control task lighting, whereas ambient lighting is typically controlled on a wide-area basis.

‡Ibid.

§http://www.toolbase.org/Technology-Inventory/Electrical-Electronics/white-LED-lighting, accessed July 30, 2009.

¶Personal communication, Joyce Kelly, lighting designer, June 2009.

**Previously the ITC had only been applicable to solar energy projects.

††http://www.acore.org/files/images/email/acore_stimulus_overview.pdf, accessed March 20, 2009.

Solar retrofits on commercial buildings, other than retail, have been hard to find included in LEED-EB projects. For most commercial applications, solar photovoltaic power is not yet economical, compared with just about anything else one can do to green an existing building. But help is on the way.

As part of the effort to incentivize the adoption of solar power in both commercial and residential buildings, the Emergency Economic Stabilization Act extended the 30 percent solar investment tax credit for eight years, through the end of 2016, for both solar thermal and solar electric (photovoltaic, or PV) systems.

This law provided much needed stability for solar investments.* Renewable energy expert Scott Sklar described the main funding mechanisms instrumental to the expanded use of renewable energy in buildings and homes:†

> The main tools are tax credits, state grants, and renewable portfolio standards. The December 2008 stimulus bill extended the 30 percent ITC for solar energy property for eight years (until 2016), with five years accelerated depreciation treatment. That is the best thing to help businesses use solar.
>
> Seventeen states have clean energy "public benefit" funds or system benefit trust funds. These are small charges on the ratepayer's electricity bill that add up to a couple of billion dollars a year that can be accessed by both homes and businesses as grants for solar installations. More than 20 states have renewable energy portfolio standards (RPS), meaning the utilities get a certain payment or rate for green power. President Obama and the Congress are proposing to make RPS a national program, [with minimum amounts of renewable energy required by 2016 or 2020.] If you blend these tools together, the payback calculation can be compelling.‡

SOLAR POWER FOR COMMERCIAL BUILDINGS

Solar power for commercial buildings still has a long payback, even with local utility incentives, but can represent a viable alternative considering non-monetary reasons, including the visibility of solar panels to the public. Alternatively, many firms have been using third-party investment partnerships to finance the installations, with the end-user buying the energy generated, typically for 15 to 25 years, at a discount from prevailing electricity rates.§ Figure 7.6 shows a typical solar power installation, in this case on the roof of the Bentley Prince Street carpet manufacturing plant in Los Angeles.

---

*Solar Energy Industries Association, "U.S. Solar Industry Year in Review 2008," www.seia.org, accessed May 26, 2009.

†Scott Sklar, president of The Stella Group, Ltd., a renewable energy strategic marketing and policy analysis firm, interview, February 11, 2009.

‡Despite the relatively large federal tax incentives and local utility rebates available, the solar payback horizon is relatively long, typically about 15 to 20 years. It is important to remember that on-site solar power generation can provide "green" marketing opportunities and can also protect against energy price volatility, blackouts, and brownouts.

§Jerry Yudelson, 2008, "Letting the Sun Shine on the Retail Sector, ICSC Research Review," available at www.greenbuildconsult.com/books/, accessed May 27, 2009.

Figure 7.6 **Solar power installations, such as this one at the Bentley Prince Street carpet manufacturing plant in Los Angeles, can be quite large, typically for commercial buildings in the hundreds of kilowatts of rated capacity.** *Courtesy of Bentley Prince Street.*

LEED-EB rewards both on-site and off-site renewable energy purchases, as shown in Table 7.8. In addition to the maximum four points shown in the table, an additional point is available for "exemplary performance," for example, by providing 13.5 percent of total energy from renewable sources, or by combining on-site generation with off-site purchases up to 100 percent of total energy use.* It turns out to be far cheaper to buy the off-site renewable energy credits than to purchase and install solar power systems on an existing building. Of course, the first step in even considering renewable energy is to cut building energy use 20 to 30 percent, to make it easier and cheaper to achieve the percentages shown in Table 7.8.

Consider the following example of a 50,000-square-foot building using 1.5 million kilowatthours per year of total energy (electricity and gas). To achieve a 12 percent contribution from solar power would require generating 180,000 kilowatthours per year. This would require a solar power system of about 120 kilowatts to 150 kilowatts, costing about $800,000. Even with a 30 percent tax credit, an accelerated depreciation

*LEED Reference Guide, p. 210.

TABLE 7.8 LEED EA CREDIT 4 POINT TOTALS*

LEED POINTS	ON-SITE RENEWABLE ENERGY, IN %	OFF-SITE RENEWABLE ENERGY CERTIFICATES, IN %
1	3.0	25.0
2	4.5	37.5
3	6.0	50.0
4	7.5	62.5
5	9.0	75.0
6	12.0	100.0

*LEED Reference Guide, p. 203. Percentages refer to percent of total building energy use, not just electricity.

deduction worth about 25 percent (net present value) and a utility payment of 30 percent of the upfront cost, the net financial cost would be about $200,000. For less than $20,000, one could purchase renewable energy certificates for 100 percent of the power from off-site sources, receiving the same number (6) of LEED credit points.*

All these considerations aside, if someone else will pay the cost of the solar power system, many retailers and others with large roof expanses have found a way to employ such systems as part of energy efficiency retrofits.† This activity was significant in 2007 and 2008, but appears to have been reduced significantly in 2009 because of the global credit crunch that has made it much more difficult for such solar investment partnerships to borrow money.

Solar for retail As an illustration of what can be done with existing buildings, it's interesting to see what has been done with existing buildings in the retail sector, especially in California and Hawaii. In late 2007, Wal-Mart announced a solar power "pilot" project in California (18 units) and Hawaii (4 units) to outfit 22 facilities with solar energy systems, with a projected output of 20 million kilowatthours (kWh) per year.‡ The company signed 10-year agreements with three companies to build, own, and operate the systems. Wal-Mart will maintain ownership of the renewable energy credits (RECs) that the solar systems produce.

In 2007, Safeway Stores, a major West Coast grocery chain, announced it would "solarize" 23 stores, mainly in California, with systems averaging about 300 kilowatts. Collectively, these systems are expected to produce 7.5 million kilowatthours per year. According to the company, "Safeway, one of California's largest renewable

*Example adapted from Brad Jones, Peter Dahl, and John Stokes, "Greening Existing Buildings with the LEED Rating System," *Journal of Green Building*, 2009, 4, 1, pp. 50–51.

†Jerry Yudelson, *Sustainable Retail Development*, Chapter 6, 2009.

‡Wal-Mart, www.walmartfacts.com/FactSheets/Solar_Power_Pilot_Project.pdf, accessed May 28, 2009.

energy purchasers, has embarked upon a major solar initiative with our solar power partners to augment our comprehensive greenhouse gas reduction initiative. There are many items that must come together to make solar economics work, with no two projects being exactly the same."* Other major U.S. retailers making similar announcements in 2007 and 2008 included Kohl's, Macy's, Target, JC Penney, Costco, and REI. As a sign that third-party solar power financing was not dead, in February 2009, the Shops at Mission Viejo (California) added a 20,000-square-foot, 173-kilowatt rooftop system costing more than $1 million and estimated to supply about 5 percent of the mall's total annual electricity consumption.†

Project Profile: The Food Bank of Western Massachusetts, Hatfield, MA
In August 2006, the Food Bank (Fig. 7.7) completed a major facility renovation and expansion, doubling its size from 17,300 to 30,096 square feet. In planning for this

Figure 7.7 **The Food Bank's 30-kilowatt photovoltaic system is part of a series of energy saving measures that contribute to reducing operating costs at the 30,000-square-foot facility.** *Courtesy of the Food Bank of Western Massachusetts, Inc.*

**Environmental Leader*, www.environmentalleader.com/2007/09/14/safeway-to-install-solar-power-panels-on-23-stores/, accessed May 28, 2009.

†www.csrwire.com/news/14498, accessed February 15, 2009.

project, the Food Bank was faced with several concerns, including the challenge of dramatically increasing refrigeration capacity without creating significantly higher operating costs. To accomplish these goals, the project:

- Upgraded all warehouse lighting to T5 lamps and all office lighting to T8s
- Provided extensive daylighting
- Installed high-efficiency (SEER 13) rooftop units for the office, with economizer "free cooling" via dual enthalpy control
- Replaced inefficient unit warehouse heaters with radiant units
- Installed five submeters for all refrigeration equipment

Many of these retrofits were supported with rebates from the local electric cooperative. In addition, the project installed an expandable 30-kilowatt PV system, supplying about 10 percent of the facility's power needs using a grant from the state. All efficiency and solar measures are expected to yield an estimated $12,000 in annual electricity savings. The project also helps serve the Food Bank's social mission by allowing more funds to be used for food purchase and distribution rather than for operating expenses.*

ECONOMICS OF SOLAR POWER IN 2009

Ron Blagus is director of energy markets for Honeywell Building Solutions, a national firm active in energy management and renewable energy. In assessing the current markets for institutional solar power installations, he says:†

> First and foremost, it's about saving money. It's about reducing the cost of operating a facility regardless of size, by making careful choices with regard to energy efficiency and renewable energy. Secondly, it's also about environmental stewardship, but environmental stewardship is tough to commit to if there's no economic benefit in doing it. Most of the entities that we deal with are not-for-profit. They have very limited budgets, which means they have very specific payback requirements for their investments. Certainly, there are environmental stewardship issues to be dealt with, but it's almost always about what sort of economic benefit is to be gained from doing this. That's where we advise customers. We help them build their strategy and then implement it.
>
> Right now the strongest incentives for any of this are probably the renewable energy credits (RECs) that are associated with solar energy. Those are market-based. The strongest markets where RECs have the most benefit are California, Arizona, North Carolina, Massachusetts, and New Jersey, and because of that, they drive the market in those states. There are a lot of states like Florida and Texas that really don't have a solar market because the cost of conventional power is low relative to the cost of power generated by solar and because there aren't economic incentives in place to

---

*Information from Food Bank of Western Massachusetts and from www.westernmassedc.com/news/foodbankofwesternmassachusettsgoesgreen_21/, accessed May 29, 2009.

†Interview with Ron Blagus, April 2009.

reward using solar energy. It's not always intuitive. California and Arizona make sense, but would you think about solar for Massachusetts and New Jersey? However, they are good markets at the moment. That will change over time. We've seen these markets emerge over the last couple of years in very unlikely places such as North Carolina and Ontario, Canada—both are very good solar markets. It's all because of this confluence of tax credits, renewable energy credits, relatively high power prices, the availability of sunshine—all of these different factors make a market.

It's hard therefore to make blanket statements about the future of solar power in LEED-EB projects. The desirability of a solar solution depends significantly on local incentives and creative financing, both of which are in flux right now.

Summary

This chapter looked specifically at energy technology choices in greening existing buildings, including building energy efficiency upgrades and adding solar power as an on-site energy system. The focus of most energy retrofits is on investments that have a demonstrable financial return and, as a result, are likely to be the first implemented. In the next three chapters, we turn to LEED-EBOM measures that have perhaps fewer tangible economic returns, but which still represent valuable investments in people's health and productivity as well as in reduced environmental impact. From a corporate sustainability perspective, both approaches are important, as we shall see in the next few chapters.

8

GREENING SITE MANAGEMENT AND REDUCING WATER USE

We've dealt with various technologies and funding mechanisms for energy retrofits at great length in Chap. 7. Now it's time to turn our attention to two other important LEED categories, Sustainable Sites, and Water Efficiency. Together, these two categories provide enough points for basic project certification (40 points), so their importance can't be overstated. Which credits does LEED-EBOM 2009 provide for site and water issues? Let's deal first with site issues. Table 8.1 shows the available credits in this broad category of site improvements and maintenance activities.

A total of 26 points can be earned in this section. If you're not in a building that was LEED-certified during original construction, there are only 22 possible points. The largest potential gain could come from reducing single-occupant commuting in conventionally fueled vehicles, accounting for up to 15 of the 22 points. The remaining measures are pretty straightforward; most projects should be able to achieve them.

Reducing Commuting by Building Occupants

LEED puts considerable emphasis on reducing total commuting by building occupants. Studies have shown that commuting can use more energy than all the building operations combined,* so in the long run, urban design is a critical green building issue. For example, for an average office building in the United States, office workers expend 33 percent more energy commuting to and from the building than is consumed by the building itself to supply them with heating, cooling, lighting, and other energy services. For an office building built to modern energy codes (such as ASHRAE 90.1-2004), auto

*http://www.buildinggreen.com/auth/article.cfm/2007/8/30/Driving-to-Green-Buildings-The-Transportation-Energy-Intensity-of-Buildings/, accessed July 10, 2009.

TABLE 8.1 LEED-EBOM 2009 CREDITS FOR SUSTAINABLE SITES (SS)

CREDIT	POINTS
1. Occupy a LEED-certified building, certified either under LEED-NC, LEED for Schools, or LEED for Core and Shell	4
2. Building Exterior and Hardscape Management Plan	1
3. Integrated Pest Management, Erosion Control and Landscape Management Plan	1
4. Alternative Commuting Transportation	3 (10% reduction in single-occupant trips) to 15 (75% reduction in such trips)
5. Site Development—maintain 25% of site area outside building with native/adapted vegetation	1
6. Stormwater Quantity Control—reduce runoff by 15% annually	1
7.1 Heat Island Effect: Non-Roof—reduce heat absorption by 50% of hardscape area	1
7.2 Heat Island Effect: Roof—put heat-reflecting roof on 75% of area or cover 50% of area with a green roof	1
8. Light Pollution Reduction—reduce lighting levels and light trespass	1

commuting uses 2.4 times the energy of building operations (36 kilowatthours/square foot of occupant space vs. 15 kilowatthours/square foot code-mandated energy use).* As energy code requirements for building energy use get more stringent, the ratio of commuting energy to building energy use may even increase, unless employers begin to focus on transportation demand management as a core part of their business operations. To qualify for points in this credit category, you'll have to survey building occupants to find out current patterns of commuting (this can be difficult in multitenant buildings). For calculating reduction in commuting, LEED assumes a baseline of 100 percent single-occupant use by regular building occupants in conventionally powered autos.†

CUTTING COMMUTING ENERGY USE

How can you cut commuting energy use? Here are some easy methods:

- Secure bicycle storage and showers/changing facilities
- Preferred parking spaces for carpools or fuel-efficient vehicles

---

*www.buildinggreen.com/press/transportation_energy_intensity.cfm, accessed May 28, 2009. Assumes an average commute of 12.2 miles.

†LEED Reference Guide, p. 23.

- Charging stations for plug-in hybrids and other electric vehicles*
- Shuttles to public transit hubs
- Compressed workweeks and telecommuting programs
- Employer subsidies for using public transit or bicycle commuting, such as free transit passes
- Purchase of hybrids or fuel-efficient vehicles for company cars that employees use. These must meet a minimum score of 40 in the annual "Green Book" of the American Council for an Energy-Efficient Economy (ACEEE)†
- Provide passes for Zipcar or other shared use vehicles so that employees can take care of personal emergencies while at work; or provide "guaranteed ride home" programs for late-working employees or those with personal crises‡

An alternative is to participate in a local or regional commute reduction program. These programs do work. For example, Stanford University is located midway between San Francisco and San Jose, California. To reduce demand for parking spaces on the campus, Stanford operates a comprehensive Transportation Demand Management program that includes formal carpool and vanpool programs, designated carpool/vanpool parking spaces, rideshare matching, assistance with commute planning, information on local transportation options, Guaranteed Ride Home program, transit/rail pass sales, promotional events, commuter checks (allowing for pretax payment of vanpool and transit/rail costs), extensive bicycle facilities/infrastructure and promotion, Clean Air Cash (non-campus residents are eligible for up to $204/year if they don't drive alone to work), and use of the Marguerite (a free campus transit/shuttle system to and from the local commuter heavy rail line).§

LEED-EB shows the way to cut commuting energy costs, which will also help lower a building's "carbon footprint." The calculated carbon footprint should include both direct and indirect effects of energy use, so reducing commute energy use is an important and often overlooked factor.¶ Table 8.2 shows how many LEED points you could get for success in this effort. For example, just getting half the building occupants on a four-day workweek will yield three points by itself, to say nothing of cutting building energy use, which should add an ENERGY STAR point or two. A 95 percent reduction in conventional commuting trips will also deliver an extra point for "exemplary performance." As higher levels of commute reduction are somewhat challenging, they are often achieved through a combination of measures. Buildings in large, densely populated urban areas with excellent public transit, such as New York, Chicago, San Francisco, and Boston, have more options when it comes to achieving

---

*If the electricity is made with renewables, this can be a good thing; otherwise, little is gained from an environmental standpoint, only the air pollution and fuel use is transferred from the car to the power plant.

†www.greenercars.org, accessed May 28, 2009. Requirement is from LEED Reference Guide, p. 25.

‡www.zipcar.com, accessed July 10, 2009.

§www.bestworkplaces.org/about/success.htm#s, accessed May 28, 2009.

¶www.carbonfootprint.com, accessed May 28, 2009.

TABLE 8.2 LEED POINTS FOR COMMUTING TRIP REDUCTION

DEMONSTRATED PERCENTAGE REDUCTION IN CONVENTIONAL COMMUTING TRIPS	SUSTAINABLE SITE CREDIT 4 LEED POINTS
10%	3
25%	7
37.5%	9
50%	11
75%	15

exemplary performance points for alternative transportation.* Or, you can just offer preferred parking at the office for carpools, hybrids, or highly fuel-efficient cars. Some employers even offer a monetary incentive for employees to purchase a hybrid or highly fuel-efficient car! Imagine what would happen if everyone who worked at your business were committed to reducing energy use through their personal transportation choices. This would be one of those "win-win" situations for the environment, as well as for employee health and well-being.

Reducing Urban Heat Island Effect with Green and Reflective Roofs

The U.S. Secretary of Energy, Steven Chu, achieved a measure of notoriety in the spring of 2009 by advocating the use of white reflective roofs and paving materials to cut energy use worldwide, which he claimed was equivalent (at 100 percent adoption) to taking all cars in the world off the road for 11 years!† You might not be able to claim such far-reaching results, but what could be easier when it's time to repair or replace your roofing than to give it a high-reflectivity coating?‡ That's the low-cost option. LEED requires a white or light-colored roof with a Solar Reflectance Index (SRI) of 78 on a low-sloped roof (less than 2:12 slope) and an SRI of 29 when the slope is greater.§ The SRI measures the ability of a surface to reflect solar heat. A black roof has a reflectance of 0.05 (5 percent), whereas a standard white roof has a reflectance of 0.80 (80 percent). The more heat that is reflected, the less comes through the roof into the working space, which in most climates will add to cooling loads.

A higher cost option with many more environmental benefits is a green roof. A green roof not only saves energy, it may even extend roof life by providing additional

*LEED Reference Guide, p. 31.

†http://ca.news.yahoo.com/s/afp/090526/usa/climate_warming_us_britain_chu, accessed July 7, 2009.

‡LEED Reference Guide, pp. 57-62.

§Ibid., p. 57.

protection against ultraviolet radiation. In addition, a green roof can act as wildlife habitat and assist with stormwater management, allowing you to claim multiple credits with a single measure. Let's explore the subject in more detail.

Green Roofs

Green roofs are an increasingly popular option for LEED-EB buildings. The City of Chicago has led the way in promoting green roofs in urban environments as a way to reduce the Urban Heat Island Effect: the documented tendency of cities to be hotter than the surrounding countryside and to require more energy use for air-conditioning in the summers.* As early as August 2006, Chicago had developed more than 200 green roofs, covering 2.5 million square feet of space, the equivalent of a 60-acre park.†

In Chap. 7, we described how the CTA Headquarters building in Chicago had installed a 29,000-square-foot green roof as part of its LEED-EB Gold certification. Figure 8.1 shows how attractive a green roof can be, especially in a dense urban environment.

Figure 8.1 The Chicago Transit Authority installed a large green roof on its headquarters building, providing a great amenity in a dense urban area. *Photo by John Herbst, courtesy of Transwestern.*

*http://www.artic.edu/webspaces/greeninitiatives/greenroofs/main_map.htm, accessed July 7, 2009.
†www.inhabitat.com/2006/08/01/chicago-green-roof-program/, accessed May 28, 2009.

Green roofs come in two varieties, intensive and extensive. An intensive green roof has more than 4 inches of soil and can support a wide variety of plants; for example, a LEED Gold residential project in New York City, the 27-story Solaire apartment building in Battery Park City, is home to a rose garden on the roof at the 19th floor.* However, intensive green roofs add more weight and require more irrigation and maintenance, so most projects use extensive green roof treatments, in which the soil layer is thinner (less than 4 inches) and typically composed of lightweight materials such as perlite.

Green roofs can also be designed as a second-floor amenity over a ground-floor retail podium in office buildings or residential high-rises in the cities. Available to office workers or residents, a green roof provides a park-like space in the midst of a city.

Green roofs are not cheap. Typical costs range from $10 to $30 per square foot. For a 15,000-square-foot green roof, that would be $150,000 to $200,000. However, for a large project with multiple stories, a green roof might represent a significant amenity at a much lower cost per square foot. In a project with high-level LEED-EB goals, a green roof can help with open-space goals, thermal comfort, stormwater management, and reducing the Urban Heat Island Effect, cutting air-conditioning costs in summer and also helping to store water from small- to medium-sized storms for future use in landscape irrigation, toilet flushing, or release to waterways.

Green roofs are used in many applications, including commercial, industrial, government, and residential buildings. In Europe they are widely used for their stormwater management and energy savings, as well as their aesthetic benefits. Green roof systems may be modular, with drainage layers, filter cloth, growing media, and plants already prepared in movable, interlocking grids. Each component of the system may be installed separately in layers. Unlike white reflective roofs, green roofs do require continuing maintenance, so they are best suited for projects with an on-site operations staff.†

In addition to contributing to Sustainable Sites credit 7.2, a green roof can be used to help secure credit 6, Stormwater Quantity Control, as well as credit 5, Protect or Restore Habitat, even in dense urban areas. Since a green roof imitates nature, this list clearly shows that it can also provide multiple ecosystem services just as one receives from natural habitats.

Creating and Maintaining Green Sites

Many of the other LEED-EB Sustainable Sites credits can be grouped in the category of creating and maintaining green sites. These include:

- Credit 2: Building Exterior and Hardscape Management Plan
- Credit 3: Integrated Pest Management, Erosion Control, and Landscape Management Plan

---

*www.thesolaire.com/documents/rooftop.html, accessed May 29, 2009.

†Jerry Yudelson, *Green Building: A to Z: Understanding the Language of Green Building*, 2007, Gabriola Island, BC: New Society Publishers, pp. 82–84.

- Credit 5: Site Development: Protect or Restore Open Habitat
- Credit 6: Stormwater Quantity Control
- Credit 7.1: Heat Island Reduction: Non-Roof
- Credit 8: Light Pollution Reduction

One of the significant issues for large buildings is stormwater management, including reducing the quantity of runoff and improving its quality. The installation of rain gardens as a measure to channel and slow down the runoff through landscaped areas is increasingly popular in the rainy Pacific Northwest, as illustrated by this example.

CASE STUDY: OREGON CONVENTION CENTER, PORTLAND

In September 2008, the Oregon Convention Center (OCC) (Fig. 8.2) was recertified with the LEED-EB Silver rating from the USGBC, making it the first convention center in the nation to achieve recertification at a higher rating. In 2004, the center was rated LEED-EB certified, upon completion of a major expansion.*

A 1 million-square-foot building with 255,000 square feet of exhibit space, the OCC is the largest facility of its type in the state of Oregon. Notable green features of the expansion include:

- A rain garden with an extensive system to take rainwater from the facility's roof and filter it through a series of settling ponds and landscaping before releasing it into the nearby Willamette River.
- The grounds and the building exteriors are both designed to reduce the "heat islands" produced by asphalt, concrete, or hard-surface roofs. The use of natural habitat vegetation in landscaping and the facility's roof design meet ENERGY STAR requirements for emissivity and reflectance.
- OCC is a top-level supporter of the Blue Sky program, Pacific Power's voluntary wind power purchase program for offsite renewable energy. The Blue Sky program is one of the three options giving Pacific Power's customers a way to help stimulate the demand for renewable energy. These purchases helped the OCC achieve all four LEED points for purchasing offsite renewable power.

In addition to meeting LEED-EB operational standards, the OCC's expansion design by architectural firm Zimmer Gunsul Frasca incorporated numerous design elements to reduce runoff and cut down on water use. Stormwater protection devices included the new rain garden. Other water-conservation features include the use of native plants, efficient irrigation technology, and a minimized lawn area.†

As the OCC example demonstrates, there are many things one can do with existing buildings to reduce environmental impacts from site use. For example, one can restructure site maintenance contracts to meet LEED-EB requirements for lower-impact

*http://www.oregoncc.org/sustainability/, accessed May 29, 2009.

†http://www.allbusiness.com/north-america/united-states-oregon/1126179-1.html

Figure 8.2 **The Oregon Convention Center takes a number of measures to reduce the Urban Heat Island Effect, including the use of both natural habitat and a rain garden in the landscaping design.** *Photo by Nancy Erz.*

management practices. Replanting with native or adapted vegetation can also reduce requirements for pesticide and herbicide use, as well as help to reduce the amount of stormwater that flows off the site. Native tree plantings could also help to gain credit for reducing the solar absorption of hardscape materials by providing shading during the hot season. The LEED-EB requirements are not difficult to meet, but do require careful attention not only in creating plans for operations and maintenance practices, but also in documenting how the project has implemented plan objectives during the performance period.

HABITAT RESTORATION

For habitat restoration, LEED requires that 25 percent of the area outside the building footprint be planted with native or adapted species (or 5 percent of the total site area, including the building footprint, whichever is greater). In addition to plantings, a project can use natural site elements such as water bodies, exposed rock, natural (unplanted) ground, and other natural elements. In the Sonoran Desert of Arizona, for example, most of the landscaping will consist of natural desert ground, with a few widely spaced trees, cacti and shrubs, some rocks and little, if any, water.

In a wetter environment, such as Sturtevant, Wisconsin, JohnsonDiversey Global Headquarters (Fig. 8.3) achieved a LEED-EB 1.0 Gold certification in 2004 with a landscape management program focusing on native plantings and stormwater management. The company added a pump into the stormwater detention ponds to irrigate the lawn and flower beds near the building. The ponds provide sediment, pollution, and stormwater control. Water conservation improvements, both inside the building and in the landscape irrigation practices, reduced overall city water use by more than 2 million gallons per year.*

Figure 8.3 **At its 277,000-square-foot global headquarters in Sturtevant, Wisconsin, JohnsonDiversey adopted many of the environmentally friendly site management practices specified by the LEED-EB program.** *Photo by Bob Israel, courtesy of JohnsonDiversey.*

*www.fmlink.com/ProfResources/Sustainability/Articles/article.cgi?USGBC:200709-22.html, accessed May 29, 2009.

What did it take to create natural habitat on this nearly 50-acre site?* JohnsonDiversey hired an outside landscape nursery to monitor the growth and health of native or adapted vegetation such as prairie grasses, wildflowers, and native landscaping. Each year, the nursery provides a landscape assessment along with improvement recommendations. Properly vegetating this site had been an ongoing issue since construction was completed in 1997. After extensive site disturbance during construction, efforts to restore native vegetation were challenged by failed plantings and difficulties with invasive species. JohnsonDiversey eventually hired a consultant with extensive knowledge of regional native species and appropriate planting and maintenance techniques, who was able to make the plantings work as intended.

Once established, the vegetation has been largely self-sustaining. JohnsonDiversey's consultant performs annual checks to ensure invasive species are not infringing on the site. JohnsonDiversey plans to closely monitor and observe the health of the native plantings and vegetation that have been introduced onto the site.

LIGHT POLLUTION REDUCTION

Light pollution reduction strategies include partially or fully shielding all exterior fixtures over 50 watts, so that they do not directly emit light to the night sky. Interior lighting should include automatic controllers to turn off lighting visible from windows at 10 p.m. or immediately after closing.† A project can also meet this credit by measuring night illumination levels at the property's perimeter and determining that all of these levels are not more than 20 percent greater than the level measured with the lights off. Nighttime illumination of landscaping and facade is acceptable with low-wattage and focused light sources.‡

Water Efficiency

LEED-EBOM awards up to 14 points for water efficiency attainments, and a project can also attain additional bonus points for exemplary performance and for meeting regionally significant water-use reduction targets. Water is emerging as the next big environmental concern for two reasons: one is the relative scarcity of freshwater resources around the world, especially in the context of global warming and increasing world population. The second reason is that the process of capturing, storing, transporting, distributing, and treating water is a large net consumer of significant amounts of electric power.§ While the majority of water use is still in agriculture and industry, urban water resources are constrained by aging infrastructure, population

---

*JohnsonDiversey case study by Stu Carron, accessed May 29, 2009, http://www.seco.cpa.state.tx.us/TEP_Production/h/green04.html

†LEED Reference Guide 2009 pp. 65–66.

‡LEED Reference Guide, p. 72.

§http://www.sandia.gov/energy-water/nexus_overview.htm, accessed July 10, 2009.

growth, and fast-rising costs. For example, wholesale water prices in southern California are projected to double between 2008 and 2013.*

WATER EFFICIENCY PREREQUISITE

Recall that LEED-EB contains one Water Efficiency category prerequisite: a project must reduce water use below the LEED baseline. What does that mean? The baseline is a calculated number that assumes that all building fixtures meet the 2006 editions of the Uniform or International Plumbing Codes.† The baseline is adjusted upward 20 percent from the code limits for buildings completed in 1993 or later and 60 percent for older buildings. For example, assume that one owns a 1990-built structure with 100 water closets flushing at 3.5 gallons per flush (gpf). The current code is 1.6 gallons per flush, and the baseline would be the equivalent of 2.56 gpf (1.6 gpf times 1.6 baseline adjustment). So you'd have to reduce water use by about 27 percent to meet the prerequisite. That means you'd have to change out roughly 50 percent of the toilets to fixtures that met the new codes (94 gallons required savings divided by 1.9 gallons per fixture savings). Typically, at today's water rates in most cities, one would make money doing this since future water savings would more than pay for the costs of the fixture upgrades.

LEED Credit Requirements

Table 8.3 shows how LEED-EBOM allocates points among various water issues. There are four credits and 14 total available points in this section. Most projects should be able to get five to seven of these points with minimal effort.

TABLE 8.3 LEED-EBOM WATER EFFICIENCY CREDITS AND POINTS

CREDIT	POINTS
1. Water Performance Measurement—Whole Building or Submetering	1 to 2
2. Indoor Plumbing Fixture and Fitting Efficiency	1 (at least 10% below baseline) to 5 (30% below baseline)
3. Water Efficient Landscaping	1 (50% below calculated baseline) to 5 (100% below baseline)
4. Cooling Tower Water Management	1 (chemical management via conductivity) and 1 (>50% non-potable makeup water)

*http://www.pe.com/localnews/inland/stories/PE_News_Local_S_cuts15.45082a4.html, accessed July 10, 2009.
†http://www.iccsafe.org/e/prodcat.html?catid=C-P-06i&pcats=ICCSafe,C,I-C-06&stateInfo=clkfdpAjbpElluan 952313, accessed July 10, 2009.

METERING AND SUBMETERING

For this credit especially, one might say, "what gets metered, gets managed." If a building isn't metered for water use, there's little hope of justifying water use reductions. Of course, for most commercial and institutional buildings (except perhaps in campus settings), there is at least a whole building meter. If one additionally provides submeters for at least 80 percent of the total water use of one major water-using subsystem such as irrigation, indoor fixtures, domestic hot water, or process loads (dishwashers, clothes washers, pools, etc.), then the project can earn two points in this section.* If cooling towers, a major water user, are submetered, then LEED requires all of the towers to be metered, to qualify for a credit point. Depending on the nature of the building, the project management team might decide to measure one or more of the subsystems. If there is gray water recovery and treatment or rainwater harvesting and reuse, then those subsystems can be monitored instead of others. Monitoring two or more subsystems can earn another point for exemplary performance.

WATER USE FOR IRRIGATION

The tie-in between the previous section on sustainable site management and reducing potable water use for irrigation should be clear. By planting native and/or adapted vegetation, a project should be able to reduce potable water use for landscaping maintenance as well as reduce pesticide/herbicide use and increase natural habitat. By calculating a baseline use for irrigation using regional averages, a project can demonstrate its level of water-use reduction. Of course, if the native plantings are well established already, it may be possible to reduce water use by 87.5 percent (4 points) or even 100 percent (5 points). Notice that LEED requirements address only potable water use (including well water), so if you're capturing and treating gray water or rainwater for reuse in irrigation, you can still meet the 100 percent reduction requirement by relying solely on those sources.†

WATER USE FOR PLUMBING FIXTURES

In commercial applications, there is a range of acceptable high-efficiency toilets (HETs) that reduce water use 20 to 30 percent from the conventional 1.6-gallons-per-flush (gpf) toilet, down to 1.28 or even 1.12 gpf. There are many brands on the market of both waterfree and low-water urinals (i.e., those using 1 to 4 pints per flush) to reduce water use 50 to 100 percent below that of the standard 1-gpf urinal. Most of the building managers we interviewed for this book prefer the low-water using urinals, shown in Fig. 8.4, reasoning that they have fewer code-compliance problems, less maintenance, and possibly fewer odor concerns. They still deliver most of the water

*LEED Reference Guide, *op. cit.*, p. 93.

†Ibid., p. 101.

Figure 8.4 **The 1-pint-per-flush urinal seems to be emerging as a strong contender to save almost 90 percent of the water used by conventional urinals, while still looking and working like a standard fixture.** *Courtesy of Zurn Industries, LLC.*

savings of a waterfree urinal: a 1-pint-per-flush urinal will save almost 90 percent of water use compared with a standard urinal. (There's a lot of fluidity in the sanitary fixture business today, so these judgments are subject to change.)

Saving 35 percent of fixture water use against the LEED baseline will yield an additional point for exemplary performance. Consider a building constructed in 1985, which still has the original conventional fixtures (such as 3.5-gpf toilets) from that period. The LEED baseline is 160 percent of current code; saving 35 percent against that baseline is equivalent to simply installing all new fixtures that meet current codes! In this case, one would do a good thing, save money and water, and get six LEED points, plus possibly another for Regional Priority credits (see Chap. 10), particularly in the more arid regions of the country. All of a sudden, you realize that water efficiency upgrades could easily represent a 20-point benefit for LEED certification and pay for themselves in almost every case.

Water audits and performance contracting There are also third-party energy service companies (ESCOs) that will pay for the entire water conservation installation and take their return from the savings on your water bill. Here's how one such company describes its approach. Mark Morello is president of Infinity Water Management based in Florida.* He says:

> We do what's called performance contracting. Everything we do has to pay for itself in savings. If we go in and do a project, the utility savings—from water and energy—pays for the project. If there's no payback, we don't have a project.
>
> We're taking toilets that have anywhere from 3.5 to 5 to 7 gallons per flush and we're putting in anywhere from a 1.6-gpf to a 1.28-gpf or a 1.12-gpf fixture, depending on the cost of the water in the area in which we're working. That gives us a little bit of flexibility in what we can put in. If we're working with a high water rate, we can use more expensive fixtures and equipment that is on the cutting edge. If the water rate is not that high, we tend to use standard handle-flush valve, 1.6-gallon toilets, for example.

For Morello, the key is to perform a detailed water-use audit, using at least three years of data if it's available, and having at least the specific information on end-uses for 95 percent of water use, to figure out exactly where water use is coming from. This is a good approach for all building and facility managers to emulate.

> We look at the bills for the past three years and we'll determine with our proprietary map where all the water is going—toilets, faucets, showers, urinals, pools, irrigation, etc. Once we determine where the water is going, we'll go in with our water-saving measures and determine how much water would be saved by changing them out with more efficient options and that gives the payback. Our payback is usually within three to five years.

Morello says the quickest payback is to put 0.5-gpm (gallon per minute) spray aerators to replace 2.5-gpm aerators. The next quickest payback in older buildings is to replace the toilets. If there is a commercial kitchen, he likes to replace water-cooled ice-makers and refrigeration/freezer units with air-cooled units to save water and improve overall energy efficiency (after all, some device has to make the chilled water!).

It's not always wine and roses Of course not everything in the water-fixture retrofit business is so easy. Wade Lange is vice president of property management for Ashforth Pacific in Portland, Oregon, which owns Liberty Centre, a 17-story Class A office tower situated in the heart of Portland's Lloyd District, with about 277,000 square feet of office space (Fig. 8.5). Lange and his staff worked hard for more than two years to get a LEED-EB v.2.0 Silver rating. As for the plumbing fixtures, Lange says:†

---

*Interview with Mark Morello, May 2009; www.infinityh2o.com, accessed May 28, 2009.

†Interview with Wade Lange, April 2009.

Figure 8.5 **Liberty Centre was completed in 1997 and achieved LEED-EB Silver in 2008. Over a 2-year period, the LEED-EB project helped cut building energy use by nearly 16 percent and water use by 9 percent.** *Courtesy of Ashforth Pacific.*

We replaced all of the [toilet] flush-values so they were low-flow. There was an investment there in materials and we did it in-house. We had to have the low-flush valves because we couldn't qualify [for the LEED prerequisite] without them. The building was built [in 1997] with low-flush valves, but they didn't work and were taken out early on. So we had to go back and put in what was originally there. The staff came up with a fix to the low-flow issue; now it works and we're saving a lot of water as a result.

We couldn't replace the original valves with low-flow valves until we figured out how to make them work properly. They wouldn't move the debris that needed to be flushed and that's why we took them out originally—they didn't work. A couple of things were happening. One was fluctuation in city water supply [pressure] and the other was just the way the system was designed. If you have more than one or two toilets flushing at the same time, there was such a pressure drop that there wasn't enough pressure to adequately flush a toilet. So the fix was actually to create a holding tank of sorts up in the penthouse and gravity-feed down to maintain a constant pressure within the system. We were able then to go back to the low-flow valves with the constant pressure. They worked; and a side benefit that we didn't realize when we were working on this project was that we run less domestic water pumps so we're saving energy at the same time. That was an idea and a fix that came out of our in-house engineering staff.

The lesson here is obvious; some of this retrofit stuff is pretty technical, and you really need to know your building. Hence the virtue of having an in-house engineering staff that is committed to tackling and solving problems such as those described above.

WATER USE IN COOLING TOWERS

Most of those energy-using chillers have an associated cooling tower that uses water evaporation as part of the process. The average water use is 3-gpm per ton of cooling.* A large commercial building with 1000 tons of refrigeration will use 3000 gallons per *minute* of water. Consider that the average household might use 300 gallons per day,† and you can see the dimensions of the problem. There is one point in LEED for supplying at least 50 percent of the cooling tower makeup water from non-potable sources such as harvested rainwater, harvested gray water, swimming pool filter flush, municipal reclaimed water (purple pipe), and other sources. If you supply at least 95 percent of cooling tower makeup water this way, there is an additional LEED point for exemplary performance. LEED also rewards careful water management that reduces the required amount of cooling tower makeup water.‡

Summary

This chapter focused on sustainable site management and water-conservation strategies, which include switching to site management practices with fewer adverse environmental impacts, adding solar-reflective or green roofs (to reduce the urban heat island effect) as well as developing alternative commuting arrangements for building occupants. The discussion showed how companies can implement a variety of green

---

*LEED Reference Guide, p. 116.

†www.epa.gov/watersense/water/why.htm, accessed May 28, 2009.

‡"Makeup water is fed into a cooling tower system to replace water lost through evaporation, wind drift, bleed-off or other causes," LEED Reference Guide, p. 500.

management practices, including practicing green building exterior and site management, integrated pest management, and shading of hardscape surfaces to reduce solar heat gain. We also looked at habitat restoration, stormwater management, and light-pollution reduction as viable strategies toward LEED-EB certification. Finally, we showed how projects can easily implement water conservation strategies that will produce 15 to 20 LEED points toward the certification goal. In Chap. 9, we'll begin looking inside the building to management practices that deal with materials and resources conservation and improving indoor environmental quality.

9

GREENING THE INSIDE OF THE BUILDING

Thus far, we've dealt with water and energy use and site management. Now it's time to look at other sustainable practices, in purchasing and waste management, and in improving the indoor environmental quality of the building. While these considerations only receive 24 percent of the available credit points in the LEED-EBOM system, they are important for implementing a company's sustainability policy, and they directly impact everyone who works in the building. LEED addresses purchasing and waste management policies directly through the fourth and fifth major credit sections, Materials and Resources, and Indoor Environmental Quality. Let's start with Materials and Resources, a section whose overriding goals are to promote sustainable purchasing and sustainable waste management practices.

Materials and Resources Credits

Table 9.1 shows the credit distribution in this section; there are 10 available points and two prerequisites that must be met by all projects. The two prerequisites address the two most obvious requirements for green operations: sustainable purchasing and sustainable waste management and disposal practices. The first deals with materials that go into the building, while the second deals with what happens to them at the end of their useful life. Applying the mantra, "all waste is food"* to the latter prerequisite, LEED seeks to promote the reuse of all materials leaving the building and their diversion from landfill disposal.

SUSTAINABLE PURCHASING

LEED-EBOM groups sustainable purchasing into five distinct areas that correspond to how most buildings are operated and managed, with different line items

*www.mbdc.com/c2c_ee.htm, accessed May 29, 2009.

TABLE 9.1 MATERIALS AND RESOURCES CREDITS (10 POINTS TOTAL)

CREDIT	POINTS
Prerequisite: Sustainable Purchasing Policy	None; required for all projects
Prerequisite: Solid Waste Management Policy	None; required for all projects
1. Sustainable Purchasing—60% of ongoing consumables	1
2. Sustainable Purchasing—40% of durable goods	1 to 2 points
3. Sustainable Purchasing—50% of purchases for facility alterations and additions	1
4. Reduce Mercury Content of Fluorescent Lamps	1
5. Sustainable Purchasing—25% of all food and beverage purchases	1
6. Solid Waste Management—conduct a Waste Stream Audit	1
7. Solid Waste Management—reuse, recycle or compost 50% of consumables waste stream	1
8. Solid Waste Management—reuse or recycle 75% of durable goods waste	1
9. Solid Waste Management—divert 70% of waste from facility alterations or additions from landfill or incineration	1

for consumables, durable goods, facility alterations and additions, lamps with mercury, and food service. To get started with this credit category, you'll need to craft a sustainable purchasing policy that deals first with consumables within the building and site management's control.* The sustainable purchasing policy is a prerequisite and needs some thought, so that it incorporates each of the LEED categories of sustainable products. Each policy must specifically address the following six items. Unlike many of the LEED credits, the sustainable purchasing policy does not require documentation of achievement to demonstrate compliant policies.†

- Scope of the policy, describing the processes to which it applies
- Performance metric(s): how performance will be measured or evaluated
- Goals (these can be "stretch" goals that you don't plan to attain right away)
- Procedures and strategies, how you intend to meet the goals of the policy
- Responsible party: who is tasked with implementing the policy and what are their major tasks
- Time period: how long will the policy apply

*LEED Reference Guide, p. 239.
†Ibid., p. xxv.

The policy for "environmentally preferable purchasing" (EPP) must also apply to one other major category of purchasing beyond consumables: durable goods, facility alterations and additions, or reduced mercury in lamps. The U.S. Environmental Protection Agency has long promoted EPP, and you can meet this prerequisite just by following the EPA's guidelines.*

Helee Hillman is a senior project manager at Jones Lang LaSalle who worked with McDonald's on securing LEED-EB Platinum certification for its headquarters building in suburban Chicago. The project received one bonus innovation point for exemplary performance in its sustainable purchasing, in addition to receiving all 16 of the points in the Materials and Resources section of LEED-EB v.2.0. According to Hillman, the LEED certification effort derived from McDonald's corporate commitment to sustainability:†

> McDonald's wanted to pursue LEED-EB because they have a commitment to sustainability in everything they do—from their real estate and their supply chain to their packaging and their employees, etc. both in the restaurants and on the corporate side. They were looking at their own buildings and their own real estate—more specifically the corporate campus in Oak Brook (Illinois) where there are four different buildings. Of the four, McDonald's did a study to determine which would be the most feasible for LEED-EB certification, based on where each building was in terms of energy, water, and sustainable site criteria. The motivation was to continue their ongoing commitment to sustainability through gaining the LEED certification.

Consumables For consumables (materials with a low unit cost that are regularly purchased such as office paper and supplies, toner cartridges, batteries, and binders), LEED awards a point if 60 percent of total purchases during the performance period meet the following criteria:

- 10 percent postconsumer content and/or 20 percent postindustrial material, such as high recycled-content copier paper, and binders
- Rechargeable batteries
- 50 percent of materials locally sourced, that is, harvested and processed or extracted within 500 miles of the project
- At least 50 percent of the content should be from Forest Stewardship Council (FSC) certified paper products‡
- Contain at least 50 percent rapidly renewable materials, such as agrifiber products, cork, and bamboo

LEED allows you to "double-count," so be sure to look for products that have more than one of these characteristics, so you can count the purchase value twice, to more

*www.epa.gov/epp, accessed May 29, 2009.
†Interview with Helee Hillman, May 2009.
‡www.fscus.org/paper, accessed May 29, 2009.

easily reach your percentage targets. Since most large companies (and many smaller ones) have centralized purchasing, it shouldn't be too hard to "tag" all of these purchases as they are made and to deliver the data for the chosen performance period. Of course, you'll need cooperation from all of the manufacturers of the consumables you purchase, but that is getting easier all the time. For example, Office Depot's Fifth Green Book™ catalog contains nearly 3000 items that will likely meet these criteria.*

One of the clients of Yudelson Associates is a very large global property management company. The corporate purchasing director told us that he planned to "tag" all of the items in the company's standard catalog for consumables and durable goods with their environmental characteristics, as a starting point. His plan initially was to encourage his site managers to use the eco-labels to guide their purchases. After some period of time, the non-ecolabeled products will disappear from the company's catalog and managers will be forced to make the environmentally preferable purchasing choice.

Durable goods LEED divides durable goods between electric-powered equipment (such as computers, monitors, copiers, printers, and the like) and furniture. Projects can earn a point if 40 percent of the purchase dollars for durable goods follow EPP guidelines. Replacing gas-powered equipment with electrically powered equipment is one criterion, while the other is to purchase ENERGY STAR-qualified equipment. For furniture, LEED awards a point if 40 percent of total purchases during the performance period (which must be at least three months' duration) contain recycled content, rapidly-renewable materials, or FSC-certified wood. Furniture salvaged on-site or off-site also qualifies as an environmentally preferable durable good.

Facility alterations and additions So far, so good, but tenant improvements, moves, adds, and changes are the lifeblood of most buildings and represent far more complex undertakings than consumable and durable goods purchases. This credit is awarded if 50 percent of all purchases meet sustainable criteria. In addition to the sustainable or environmentally preferable purchasing criteria outlined above, LEED-EBOM uses the same criteria as LEED-NC for carpet and carpet cushion, resilient flooring (non-carpeted), adhesives and sealants, paints and coatings, and composite wood and agrifiber products.† For large organizations using outside designers and contractors, you'll have to make sure that they use "green specifications" in their standard approach to remodels and additions.

Reduced mercury in lamps "Houston, we have a problem!" The most popular form of relamping, using lower-wattage linear or compact fluorescent lamps, has the potential to create a hazardous waste problem if there is breakage. For a long time, mercury has been recognized as a source of environmental contamination and as a human health hazard.‡ In fact, coal-fired power plants are the source of about 40 percent

*www.community.officedepot.com/environment.asp, accessed May 29, 2009.

†LEED Reference Guide, pp. 269–281.

‡www.epa.gov/hg/health.htm, accessed May 30, 2009.

of the mercury escaping into the environment each year.* Purchasing low-mercury content lamps, with a long life and high light output is the best way to reduce the overall amount of mercury in the environment.†

How does a facility manager or building manager go about reducing mercury content in lamps? The first step is easy: inventory what you have, then start buying reduced-mercury-content lamps. LEED gives three basic measurements:

- Mercury content (mg of Hg—mercury—per lamp)
- Light output per lamp in lumens (at 40 percent of rated lifetime)
- Lifetime in hours

The LEED requirement is that 90 percent of lamps purchased during the performance period have average mercury content of no more than 90 picograms per lumen-hour (pg/lumen-h). A picogram is one million-millionth of a gram or one-millionth of a microgram. (If 90 percent of the lamp purchases average 70 picograms per lumen-hour or less,‡ LEED awards another point for exemplary performance.) Mercury content is measured for all indoor and outdoor mercury-containing light fixtures, including both hardwired and portable fixtures, such as task lamps.§ Compact fluorescent lamps that meet industry guidelines for maximum mercury content can be excluded from the calculation.

Most manufacturers will supply this information so that the calculation isn't too onerous. After all, their goal is to sell you lamps. For example, Osram Sylvania's Web site provides a LEED-EBOM calculator for mercury content.¶

Sometimes the bigger challenge is getting a building or facility's maintenance staff to become familiar and comfortable with using the newer lamps. Doug Norwood at the Sacramento Municipal Utility District related this experience:**

> Once we started getting into that credit, we figured out what we wanted to use and created a great process to do so. Well the process that we put into place didn't match up very well with the crews that actually replace them. We put in a process where we would purchase them and we would warehouse the correct lamps for each building. Then when the crews needed to replace a lamp, they would just get it from the warehouse. Well, that wasn't the way that they operated. When they needed to replace a

---

*www.sylvania.com/cgi-bin/msmgo.exe?grab_id=100&page_id=13373184&query=mercury&scope = searchost&hiword=mercur%c3%89%20mercur%20mercuric%20mercurio%20mercuryis%20mercuryor%20mercurys%20mercury, accessed May 30, 2009.

†Ibid., p. 284.

‡The average 4-foot linear fluorescent lamp in 2003 contained about 10 milligrams of mercury. If the average lifetime is 25,000 hours and the average light output is 3000 lumens, then the mercury content is simply $(10 \times 10^{-3}/3 \times 10^{3} \times 0.025 \times 10^{6}) = 133$ pg/lumen-hour, about 50 percent above the LEED limit of 90. As manufacturers are continually reducing the mercury content of lamps, increasing light output and extending lifetime hours, everything is moving in the right direction.

§Ibid., p. 283.

¶www.sylvania.com/cgi-bin/msmgo.exe?grab_id=100&page_id=11865856&query=leed%20eb&scope=searchhost&hiword=eb%20leed%20leedh, accessed May 30, 2009.

**Interview with Doug Norwood, March 2009.

> lamp they went to their storage cabinets. They would pull out what they've stocked in their storage cabinets, which they get from wherever they want to get it. They may run down to Home Depot and get them. (Actually, that's one of the things that they had done.) In just about every janitor closet, they had purchased and stored a huge inventory of fluorescent lamps that did not meet the low-mercury standards. That's where they wanted to continue to go for replacements and they didn't want to just dispose of all of those because they had made a pretty significant investment in them.

As with many of the LEED credits that require some changes in behavior, not only is considerable time required in user or occupant education, but also someone has to stay on top of the situation to monitor compliance with new policies and procedures.

Food Ever wondered where your food actually comes from? Most of us are supporting a very long pipeline of food, by some estimates more than 1500 miles for the average meal, from farm to plate.* The "slow food" movement has grown for the past 20 years as a reaction to both fast food and long food pipelines. In this decade, many food service operations, particularly on college campuses, have switched to an emphasis on organic, locally grown food.

If a building or facility contains a food service operation, LEED provides an opportunity to take sustainable purchasing into this realm as well. LEED requires that at least 25 percent of total combined food and beverage purchases during the performance period meet one or both of the following criteria: produced with a 100-mile radius of the site and be labeled USDA Organic, Food Alliance Certified, Rainforest Alliance Certified, Protected Harvest Certified, Fair Trade, or Marine Stewardship Council's Blue Eco-Label.† If food purchases are both local and certified, you can double-count the amounts to reach the 25 percent threshold.

This performance goal may be harder to achieve than it sounds. I like to think of myself as an environmentally aware consumer. One Saturday in May, I went to the local farmer's market and bought produce, fruit and eggs, but my lunch meal of a soybean-based veggie burger and potato salad probably came from more than 500 miles away from my home in Arizona and wasn't certified by any LEED-recognized standard. (I'm sure the Diet Pepsi was locally bottled, however.) Now consider the difficulty of getting an entire organization to adhere to sustainable food purchasing standards and you'll see why achieving this particular LEED credit might be quite challenging!

SUSTAINABLE WASTE MANAGEMENT

The solid waste management policy prerequisite required by LEED-EBOM covers the waste streams that are within the building or site management's control, including ongoing consumables, durable goods, and facility alterations and additions. In the second half of the Materials and Resources section, LEED provides credits for actual performance in implementing these policies.

---

*www.worldwatch.org/node/827, accessed May 30, 2009.

†LEED Reference Guide, p. 291.

Waste audits The first credit in this section is for conducting a waste audit of the building or site's consumables waste stream during the performance period, to identify opportunities for increased recycling and waste diversion. This practice can be quite revealing: most of the waste stream may consist of paper that can be easily recycled before it gets to the dumpster. The waste stream audit must analyze the contents of both what is sent to landfill and what is recycled, reused, or composted on-site.*

Theresa Townsend talks about the requirement for a waste audit at the State of California's Department of Education building:†

> The most surprising thing was [the prerequisite for] a waste characterization. We dubbed it fondly as "the dumpster dive." It requires you to actually hold your garbage for one night, go through it and characterize it so you can see what type of waste the building occupants throw out each evening. It's very telling and it really gave us some good ideas of how we could divert waste.

Composting should not be neglected, especially if there is a large food-service operation on-site. Not only does a building save money by using vermiculture (worm composting), but it may also be able to bag and sell the compost for use in home gardening. Craig Sheehy led the LEED-EB Platinum effort at the 950,000-square-foot California EPA building in Sacramento while at Thomas Properties Group; 10 tons of waste from

Figure 9.1 The Vision Service Plan headquarters in Rancho Cordova, California, has an active recycling program. *Photo courtesy of VSP/Rancho Cordova, CA.*

---

*Ibid., p. 300.

†Interview with Theresa Townsend, March 2009.

the food service operation from all the state workers' meals was composted, bagged, and sold as "Bureau-Crap," saving $10,000 per year in waste disposal costs.*

Ongoing consumables To receive a LEED credit in this category, a project must reuse, recycle, or compost 50 percent of the ongoing consumables waste stream during the performance period, measured by weight or volume. The waste stream includes paper, toner cartridges, glass, plastics, cardboard and corrugated, food waste, and metals. The project must also divert 80 percent of discarded batteries from the trash.†

Some organizations that serve the public, such as the Oregon Convention Center, have made their recycling programs into a major marketing feature. Recertified at LEED-EB Silver in 2008, the OCC says that it "manages an extensive material and waste recycling and recovery program that includes pre- and postconsumer organic waste, cardboard, newspaper, cans, plastics, glass bottles, wood pallets, cooking oil, and landscaping trimmings." It diverted 266 tons of materials from landfills in fiscal year 2007 to 2008, with an overall diversion rate of 43.6 percent (up about 12 percent from the previous fiscal year) and hopes to achieve a 50 percent rate in 2008 to 2009.

Almost every organization can use their waste recycling programs to not only save money in annual waste disposal costs, but also as a major component of corporate sustainability programs. Recycling is a practice to which every stakeholder can relate, both at work and at home. Yet there can be problems finding vendors who will take all of the LEED-required waste streams. At one project in a major city, the property manager was unable to find anyone to pick up the five components of recycling mandated by the LEED system.‡

Instituting commercial recycling can have internal challenges in many settings and often requires a wider cultural change. At Caterpillar Financial, Steve Zanolini says:§

> We didn't do any recycling before and we learned over time through different initiatives that in order to get people's buy-in to recycle, you have to make it convenient. We even got to the point—this was even after we submitted the application—we actually put a recycling bin with a small 6-by-5-inch plastic, desk-side trash can to make it as easy and convenient to recycle as possible. When we did that, it really took off. It's really eye-opening [for most people] when you see this little 6-inch trash can and on the outside of it, a sign says, "This is all the garbage I can make." Everybody has gotten accustomed to separating trash. The cultural change has just been remarkable.

Durable goods LEED defines durable goods as office equipment, appliances, televisions, external power adapters, and other audiovisual equipment, and it requires for this credit, reusing or recycling 75 percent of the durable goods waste stream leaving the building, calculated by weight, volume, or replacement value. Since there are

---

*Personal communication, Craig Sheehy. See also the case study of the project, found at http://www.thomaspropertiesgroup.com/about_us/sustainability.htm, accessed May 30, 2009.

†LEED Reference Guide, p. 305.

‡Interview with Adam Rose, April 2009.

§Interview with Steve Zanolini, May 2009.

many local recycling centers for electronic waste throughout the country, achieving this percentage of recycling or reuse shouldn't present great difficulties for most organizations.*

Facility alterations and additions For this credit, LEED requires diversion of at least 70 percent of waste, by volume, generated by tenant improvements, repairs, renovations, upgrades, and additions, from disposal to landfills or incineration facilities.† This sounds easy: just put it in the tenant requirements and the management company's standard specifications. But in practice, "stuff happens." Doug Norwood of the Sacramento (CA) Municipal Utility District, SMUD, explains what happened in his case:

> An additional challenge that stands out involved the construction waste and demolition recycling credit. We wrote a policy, we put it in our specifications and then right in the middle of our performance period, we had a small construction project going on and the contractor took everything that he demolished out of the room and he threw it in the trash bins and off it went. We actually lost that point because what we didn't think about was, *even though we put it in our specifications, you have to have somebody riding the contractors and staying right on top of them* to make sure that they follow the required procedures. You have to consider the logistics of where they are going to store that material, where they are going to sort it and how that is going to be handled. Because nobody was watching it, over just a period of a couple of days, it was out and gone before we knew it. We were not able to backtrack on that one.‡

The requirements of this credit apply only to base building elements attached to the building itself, including wall studs, windows, doors, insulation, panels, drywall, ceiling panels, carpet, and other flooring material, but not furniture, fixtures, and equipment. Mechanical, electrical, and plumbing equipment, along with elevators, are also excluded from the recycling requirement.§

Sometimes it's possible to find creative uses for base building materials that would otherwise wind up in a landfill. Johnathan Sitzlar with the Federal GSA tells what happened at the Duncan Federal Building in Knoxville, Tennessee:¶

> We had some opportunities between 2005 to 2007, when we received the LEED-EB designation, with tenants who were modifying their spaces. During those [remodel] projects we put in motion sensors, retrofitted the lighting and installed recycled-content carpet. One of our tenants, the Social Security Administration, was moving to a new facility. Their carpet wasn't very old—three years at most—and did not have a lot of wear and tear, so it was still in good condition. We diverted it from the landfill

---

*As an example, the California Integrated Waste Management Board runs a very good Web site that allows you to find places to recycle just about every type of e-waste, www.ciwmb.ca.gov/Electronics/Collection/RecyclerSearch.aspx, accessed May 30, 2009.

†LEED Reference Guide, p. 317.

‡Interview with Doug Norwood, March 2009.

§LEED Reference Guide, op.cit.

¶Interview with Johnathan Sitzlar, April 2009.

> and put it through a cleaning process. The Knox County School System obtained it at a much lower cost [compared to new carpet] that saved the school system about $25,000. That kept 12,000 square feet of carpet out of the landfill.*

Indoor Environmental Quality Credits

According to a study in the 1980s by the California Air Resources Board, most Californians spent close to 90 percent of their time indoors.[†] Think about it: on an average workday, you probably got up, ate breakfast, went to the garage, got in the car, went to the office or class or factory, ate lunch indoors, drove (or were driven) home, popped open a cool drink, ate dinner, watched TV, and went to bed. Even if you worked out or shopped, it was probably indoors. In a very hot or very cold climate, spending most of your time indoors this way might go on for months! So, without question, the quality of the indoor environment is a key influence on your overall physical and mental health, as well as personal productivity.

Table 9.2 shows the LEED-EBOM credits for Indoor Environmental Quality, a total of 15 possible points grouped into three major categories: indoor air quality, occupant comfort, and green cleaning. These categories deal with the physical makeup of the indoor environment, source control, psychological factors, and ongoing maintenance practices.

INDOOR AIR QUALITY

Indoor air quality is an indisputable component of personal health and productivity. The prerequisite for any LEED-EB project is to have at least 10 cubic feet per minute (cfm) per person of outside air supply under normal operating conditions or to meet the applicable ASHRAE standard 62.1-2007.[‡] Each air-handling unit (AHU) must meet this criterion. This can often be expensive if a building has many AHUs. Sometimes for older buildings, there must be a capital investment before they can even come close to meeting this prerequisite.

Paul Goldsmith is an experienced LEED-EB consultant. He sees the minimum ventilation requirements as one of the more difficult areas for older buildings.[§]

> Getting to the energy level for the ENERGY STAR rating for the prerequisite of 69 points is a push and to meet the minimum ventilation requirements of ASHRAE 62.1-2007 is a push also. If an older building hasn't done that, then they can't get out of the gate [for a LEED-EB rating].

---

*The Building Materials Reuse Association offers an online, interactive map (http://www.bmra.org/listings/directory-map) of building material reuse resources, such as demolition and reuse contractors and recycled building material collection centers.

[†]http://eetd.lbl.gov/ie/viaq/v_voc_1.html, accessed May 30, 2009.

[‡]Ibid., p. 329.

[§]Interview with Paul Goldsmith, March 2009.

TABLE 9.2 LEED CREDITS FOR INDOOR ENVIRONMENTAL QUALITY (EQ)*

CREDIT	POINTS
Prerequisite: Minimum Indoor Air Quality (IAQ) Performance	None; required for all projects
Prerequisite: Environmental Tobacco Smoke Control	None; required for all projects
Prerequisite: Green Cleaning Policy	None; required for all projects
1.1 Indoor Air Quality Management Program	1
1.2 Outdoor Air Delivery Monitoring	1
1.3 Increased Ventilation	1
1.4 Reduce Particulates in Air Distribution	1
1.5 IAQ Management for Facility Additions and Alterations	1
2.1 Occupant Comfort Survey	1
2.2 Controllability of Systems (Lighting)	1
2.3 Thermal Comfort Monitoring	1
2.4 Daylight and Views	1
3.1 High-performance Green Cleaning Program	1
3.2 Custodial Effectiveness Assessment	1
3.3 Purchase of Sustainable Cleaning Products and Materials	1
3.4 Sustainable Cleaning Equipment	1
3.5 Indoor Chemical and Pollutant Source Control	1
3.6 Indoor Integrated Pest Management	1

*LEED Reference Guide, p. 328.

In a good way, LEED-EBOM forces a building or facility manager to evaluate all of the key operational parameters. If a project can't meet a basic prerequisite for indoor air quality, not only can it not be labeled as a "green" building, but also that failure should be a red flag for the building management to conduct further investigations. Mark Bettin at the Merchandise Mart in Chicago had a similar experience; recall that this is a 70-year old building that has been through many rounds of renovations in its history. He says:*

> One thing that was the most difficult for us to achieve for certification was actually one of the prerequisites, believe it or not. When it came down to achieving the certification,

*Interview with Mark Bettin, March 2009.

> this prerequisite [for minimum outside air ventilation] was one of the last things we achieved. That was probably just a result of the order in which things were reviewed. We had to create a baseline for the percentage of outside air that our systems bring in. We had an air-balancer come through and take readings on all of our equipment. Then a consultant did the calculations on how each piece of equipment operates and how much outside air each unit brings in. Getting the data was part of it but, because of the certification, we also had to establish a capital program for how we categorize repairs.

The second indoor air quality (IAQ) prerequisite, environmental tobacco smoke control, should be a non-issue for most facilities in the United States and Canada, simply by having a no-smoking policy inside the building or facility and requiring any smoking to take place more than 25 feet from building entrances and air intakes. Signage and special facilities for smokers that follow specified criteria are approaches for meeting this prerequisite.

Indoor air quality best management practices The first LEED credit is for developing and implementing on a continuing basis an IAQ management program based on a U.S. Environmental Protection Agency protocol, looking at strategies for controlling all possible sources of indoor air pollutants.*

The second credit is for monitoring the minimum required outdoor airflow rate, for at least 80 percent of the total building outdoor intake airflow serving occupied spaces. In densely occupied spaces such as conference rooms, a carbon dioxide sensor must be used to adjust outdoor air intake based on occupancy.† This is an example of demand-controlled ventilation, a practice used by many mechanical engineers to assure adequate outdoor air delivery.

LEED also supplies a credit for increasing ventilation at least 30 percent above the minimum outdoor air intake rates established in the first EQ prerequisite. In older buildings, this credit may be quite difficult to get. In certain climate zones, there may be an "energy penalty" for having to bring in more outside air, for example very hot outside air in places such as Phoenix or Las Vegas, and cool it to the building level. In other climates like the Southeast, where the air is hot and humid, there will be more energy used in dehumidification of incoming air at higher ventilation rates.

Filtering outside air is a good idea to reduce particulates in air distribution. The minimum efficiency reporting value (MERV) specifies how much particulate matter a filter will remove. In the fourth credit, LEED requires a MERV 13 filter for all outside air intakes and inside air recirculation returns.‡ The standard office building uses a MERV 8 or MERV 11 filter that removes 95 percent of particles from 3 to 10 microns in diameter. (By contrast, the diameter of a human hair averages about 25 microns.)§ MERV 13 filters remove 98 to 99 percent of particles above 0.3 to 1.0 micron in diameter, so it's a much "tighter" filter. In this case also, there is an "energy penalty" by

---

*LEED Reference Guide, p. 355.

†Ibid., p. 361.

‡Ibid., p. 379.

§http://hypertextbook.com/facts/1999/BrianLey.shtml, accessed May 30, 2009.

having to push building air through a much denser filter (in engineering language, there's a greater static pressure drop across the filter).*

Finally, we come to the last credit in this section, dealing with standards for IAQ management during facility alterations and additions. Basically, LEED wants you to follow a national standard from the Sheet Metal and Air-Conditioning Contractors National Association (with the picturesque acronym, SMACNA) and to flush out contaminants for a specific period before occupancy.†

OCCUPANT COMFORT

Will people ever be satisfied? Most of us have been in buildings where some of the people think it's too hot and some think it's too cold, all at the same time. LEED recognizes that occupant comfort is a key component of getting the results from a green building that most companies really want—greater productivity, improved health, and higher morale.

The first credit in this section is a simple one: implement an occupant comfort survey (and a complaint response system) during the performance period to collect anonymous responses to such issues as IAQ, thermal comfort, acoustic comfort, lighting levels, and building cleanliness. You'll have to get survey responses from at least 30 percent of the regular building occupants, which can be a daunting task, as anyone knows who does survey research, and you'll have to put a complaint response system in place. Fortunately, most well-run buildings and facilities already have such systems.‡ The goal, according to ASHRAE, is to satisfy at least 80 percent of the occupants in the space.§

The LEED-EB Platinum-rated Vision Service Plan headquarters in Rancho Cordova, California, has this approach to handling complaints about thermal comfort.

> Every complaint is logged in the computer and the system is checked for performance accuracy. Then, the temperature is measured in the space where the complaint was made. If the temperatures are within the set points, we advise the complainant to take appropriate action (add a sweater, etc.). If the temperature is not correct, we troubleshoot the system to determine the problem.
>
> To achieve this credit point, the team conducted a thermal comfort survey of employees as required by ASHRAE Standard 55. The results? Our proactive approach paid off; we achieved an 84% satisfaction rating.¶

To meet the next credit, at least 50 percent of the occupants will have to have adjustable lighting controls. LEED says that you can meet this credit by supplying

*https://wpb1.webproductionsinc.com/danforthfilter/secure/store/HEPA-Filters-MERV.asp, accessed May 30, 2009.

†LEED Reference Guide, p. 383.

‡Ibid., pp. 391–393.

‡Ibid., p. 392.

¶L. Kilmer, M. Roskoski, C. S. Leon, "Operational Excellence," in *High-Performance Buildings*, Spring 2009, p. 33, available from www.ashrae.org.

"simple switch controls to increase occupants' control over lighting," which might simply mean task lighting at each workstation. By providing 95 percent of the individual workstations and multioccupant spaces with individual lighting controls, you can earn an additional point in this credit category.*

Thermal comfort monitoring is the third credit that deals with occupant comfort. To meet these requirements, you'll have to continuously track and optimize systems that regulate indoor comfort conditions in regularly occupied spaces. You'll have to test for air temperature and relative humidity, as well as air speed and radiant temperature, the main determinants of thermal comfort. Typically a building automation system with the right number of sensors adequately distributed throughout the building will meet the requirements of this credit.†

What good is a thermally comfortable workspace if you can't see outdoors or work under natural light conditions? LEED gives a credit in this case if at least 50 percent of the regularly occupied spaces have a daylight level of at least 25 foot-candles (an archaic but still used measurement of luminance, in this case originally representing the light of 25 candles at 1 foot distance, but now defined as 1 lumen per square foot).‡ Daylighting can be achieved through windows, skylights, roof monitors, light shelves that reflect sunlight from the windows deep into the room, and other methods. The second way to achieve this credit is to provide at least 45 percent of all regularly occupied areas, a direct line of sight through windows to the outdoors. (If you don't have an office along the building perimeter, it's OK to look through the managers' office windows to see outdoors.)§ If you can achieve at least 75 percent daylighting and 90 percent views, you can get an additional credit point.¶

Thermal comfort monitoring, daylighting, adjustable lighting controls, and other measures to increase occupant comfort can have a strong influence on increasing productivity. At the SMUD Customer Service Center in Sacramento, California, the prospect of these benefits also helped sell the LEED-EB project to management, according to project leader Doug Norwood:**

> There were a few pieces of information we provided that helped our manager make that decision [to go forward with the LEED-EB project]. One of them was productivity. It is really difficult to put a number to productivity because you're really guessing, but even with very small percentages of productivity increases, the numbers add up really fast due to salaries. You've got almost 700 people working in that building, and just a 1 percent increase in their productivity would pay for the project in three months.
>
> Being familiar with the indoor environmental quality credits, we knew going into it that if we concentrated on those, we would be able to correct some of the problems that we had in the building. We had a few areas where we had a high number of complaints—cold calls, hot calls, stuffiness. Anytime you get that, especially when it's coming from

*Ibid., pp. 398, 401.
†Ibid, p. 405.
‡Ibid., p. 496.
§Ibid., pp. 409–411.
¶Ibid., p. 423
**Interview with Doug Norwood, March 2009.

> the same areas all the time, you know it's affecting people's productivity because they're not comfortable. I don't have any specific numbers, but qualitatively, I'd have to say that the complaints from the high-density areas in that building have dropped off considerably. That tells me that we've impacted productivity in a very positive way.

GREEN CLEANING

The importance of green cleaning practices can't be overstated. The one person largely responsible for making sure that LEED has incorporated green cleaning in its requirements is Stephen Ashkin, a nationally recognized expert in the subject, who defines green cleaning as an approach that protects health without harming the environment. In his words, "Green cleaning, especially as it is incorporated into LEED, is about examining the entire process of cleaning, identifying the areas that can be improved from an occupant health and environmental perspective, developing a plan and the procedures to implement the plan, executing the plan, and measuring the results."*

Ashkin has identified three main obstacles to implementing green cleaning programs in any facility: resistance to change, training, and cost/performance issues with new products and approaches.†

> The biggest obstacle in setting up a green cleaning program is just the resistance to change. In an existing building, it's already being cleaned, and if occupants aren't complaining and cost is on budget, then there must be another powerful reason to change. Buildings either have an outsourced janitorial service provider or are being cleaned with an in-house cleaning department using the organization's own people. And both have exactly the same issues when converting. Green cleaning requires helping them understand the necessary changes relative to the selection of products, including chemicals, janitorial paper, plastic can liners and cleaning equipment; along with training workers on cleaning procedures focused on protecting occupant health and reducing environmental impacts including the energy, water, and waste associated with cleaning.
>
> The challenge in creating a healthy building in many respects is not the products that are being used, but rather [to ensure] that cleaning personnel are doing the right thing. Training becomes an incredible challenge, especially when we appreciate that many of them are part-time workers or have English as their second language. Unfortunately in some buildings, cleaning budgets are so tight that there are little time and resources available for training workers, regardless of OSHA and other legal requirements. As a result [of communications issues], getting them to change is a lot more complicated than deciding to use recycled paper in copier machines because changing paper does not require retraining all the people using the machine.
>
> Cost and performance concerns are still obstacles. Years ago when we tried implementing green cleaning programs, either the products really did cost more than their conventional counterparts or at the same costs they simply didn't work as well. Today, however, that has changed and this obstacle is largely a myth. The technologies have

---

*www.ashkingroup.com/green.html, accessed May 30, 2009.

†Interview with Stephen Ashkin, March 2009.

improved significantly over the past few years and have also resulted in more competitive costs. The price and performance issue has really gone away.

LEED-EBOM has a prerequisite that a building or facility has to have a green cleaning policy that addresses the following items:*

- Purchase of sustainable cleaning products
- Purchase of sustainable cleaning equipment
- Establish standard operating procedures for green cleaning
- Strategies to promote and improve hand hygiene
- Guidelines for safe handling and storage of cleaning chemicals
- Requirements for staffing and training of maintenance personnel
- Collecting occupant feedback on cleanliness of the building or facility

The six individual credit points address implementation of the policy items in the areas of high-performance cleaning, custodial effectiveness, purchasing sustainable cleaning products and materials, purchasing sustainable cleaning equipment, the proper selection

Figure 9.2 **The Wellington Webb building, owned by the City and County of Denver and managed by Transwestern, was certified LEED Gold in 2007. The project received all of the LEED green cleaning points as part of its certification effort.** *Photography by John Herbst, courtesy of Transwestern.*

*LEED Reference Guide, p. 347.

and cleaning of entryway matting to prevent dirt from entering the building, and having an effective indoor integrated pest management program that avoids wherever possible the use of toxic insecticides and other chemicals in favor of physical controls such as cleanliness. In addition, there are other credits in LEED-EBOM that cleaning personnel can contribute to such as recycling and waste management (MR Credits 6 & 7) and exterior building and site management (SS Credit 3).

Ashkin realizes that establishing green cleaning is an ongoing process for most building owners and managers. He stays focused on the real purpose of green cleaning, which is to protect people through cleaning as a preventive public health measure.*

Figure 9.3 **The Ecology & Environment (E&E) Buffalo Corporate Center in the town of Lancaster, New York, captured 34 of the 38 possible points in these two sections, on its way to a LEED-EB Platinum certification in 2008.** *Courtesy of Ecology & Environment, Inc.*

*Interview with Stephen Ashkin, March 2009.

We can feel very comfortable about the whole concept of prevention. If we reduce the amount of "bad" stuff in a building, we would expect people to perform better and have better health because that really is a basic premise of public health—using preventive measures. In some respects, when we talk about green cleaning, it really is a low-cost health intervention strategy. If we didn't clean, people would get sick. We clean to remove the bad stuff, things that happen naturally. It's not like we clean just to keep the floors shiny. Bacteria, viruses, mold, soils, dust, dust contaminated with chemicals and heavy metals, and those kinds of things occur naturally in our buildings and in our environment. The goal in cleaning is to identify what the soils are, to remove them from the building and dispose of them properly. That's why we clean.

Summary

This chapter focused on LEED-EB measures that directly affect indoor environmental quality, particularly indoor air quality and occupant comfort and health. LEED-EB provides specific guidance on improving daylighting and views, introducing policies for IAQ maintenance, and using low-impact materials in building cleaning. While it only consists of 15 points out of 100, this section of LEED has the most direct impact on human comfort, health and performance, the real long-term goals of greening existing buildings.

How can you get started with your LEED-EBOM program? Al Skodowski of Transwestern, already a veteran of a number of LEED certifications, offers this perspective:*

Overall the program does present some challenges. You have to be committed. I always tell people, "LEED-EBOM should not be your starting point. LEED-EBOM and the certification should be your end game. Start slowly. Take [it in] small pieces. In most cases, you will find that you have already implemented many of the things in the program. You are cleaning your buildings, taking out the trash, changing light bulbs, changing filters. Those things are currently in place. They may not fully align with the LEED requirements, but it's a good start. It's a matter of biting off small [enough] pieces. Convert your [building to] green cleaning. Then convert your chemicals and supplies [to green products]. Then convert your equipment. Work through the [individual] pieces and have LEED-EBOM become the end result."

*Interview with Al Skodowski, March 2009.

10

LEED CERTIFICATION CHALLENGES AND APPROACHES

This chapter addresses the specific issues associated with managing successful certification projects, including creating the project team, engaging with the commissioning and certification process, conducting team meetings, tracking team progress, and justifying economic costs and benefits. Leading consultants provide a number of tips for getting EBOM projects done on an accelerated timetable. The company or organization needs to change purchasing and operating practices, and change its specifications for remodels and renovations. Approaches also differ between government and private sector projects.

Let's begin with the project team. Unlike the other main LEED certification systems—New Construction, Core and Shell, Commercial Interiors, and Schools—LEED-EBOM does not typically have an architect as the leader of the project team. Instead, the team leader is likely to be internal to the organization, with consultants used for things like retro-Cx, energy engineering, and green cleaning and site management. In one case, a team even said that its janitorial service provided the most helpful documentation.*

Keys to a Successful Project

There are three keys to a successful LEED-EBOM project:

- Getting organizational commitment to the process and managing the changes in the organization required for a successful project, such as changing people's recycling behavior
- Performing the required upgrades to building or facility operations, including adopting the LEED-required policies
- Managing the LEED-certification process itself, including collecting all required documentation from the many project participants

*Sebesta Blomberg project team interview, March 2009. See Chapter 13 for more details on this project.

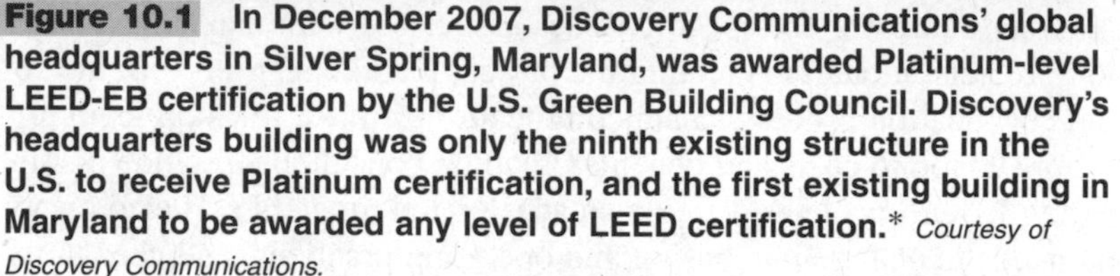

Figure 10.1 In December 2007, Discovery Communications' global headquarters in Silver Spring, Maryland, was awarded Platinum-level LEED-EB certification by the U.S. Green Building Council. Discovery's headquarters building was only the ninth existing structure in the U.S. to receive Platinum certification, and the first existing building in Maryland to be awarded any level of LEED certification.[*] *Courtesy of Discovery Communications.*

Gaining organizational commitment might be easy or difficult, depending on whether the initiative is coming from the top down or the bottom up. Either approach can work, depending on the nature of the organization. Some large property managers, such as CB Richard Ellis, Jones Lang LaSalle, Transwestern, and Cushman & Wakefield, bring the benefits of LEED-EB to their clients and then help to complete the projects. In other cases, the organization may have an exiting sustainability commitment, but it often takes a strong in-house champion to convince management to fund a project.

Sometimes the facility management team wants to make improvements, but doesn't know how to sell it to upper management or even how to work with the LEED consultant. For example, Darren Goody, commissioning manager of Portland, Oregon's Green Building Services consulting firm, has worked on more than 30 LEED-EB projects. He says:[†]

> The facility team can really engage in the process and embrace it, and they can provide information in a roadmap to help get to the measures and the findings and determine things that should be implemented right away. But there are some teams that don't want somebody else coming in. Part of that might be a little bit of fear that they are going to be exposed that they haven't done their job. So you need to be professional.

---

*http://corporate.discovery.com/discovery-news/discovery-communications-awarded-highest-level-cer/, accessed May 31, 2009.

†Interview with Darren Goody, March 2009.

> You can't come in with a condescending attitude. You need to come in with an attitude that you're a member of the team and you want to elevate the team and improve the building performance. From a people-perspective, it's about recognizing that the facility team has a lot to contribute. They will often times know what should be done, but they just haven't been able to quantify the benefit so they can't convince the building owners that they should be doing that.

This sounds like the primary issue in any change initiative: getting people to work together. *Ultimately, LEED-EB is about changing behavior as much as it is about changing light bulbs.*

With management, there might also be a perception that the cost will vastly outweigh the benefits. There may be concern about financing the required improvements, particularly if the retro-Cx project turns up ventilation issues. There may be concern about the burden on staff time, when staffs are already stretched pretty thin.

Many of the consultants interviewed for this book discount the notion that you have to spend a lot of money, particularly on any building built to Class A standards in the past 15 years. But they do acknowledge that there is a lot of fear among owners that the process will be costly. Al Skodowski of Transwestern has a work-around to deal with this perception. He says:*

> The term "LEED," to those unfamiliar with it, tends to worry people. If you mention LEED to someone they may very well go to their computer, Google it, and say, "How on earth am I going to handle all this paperwork." It is a process. Take it slowly. Begin applying some of the practices, and as people get comfortable with those, start to make LEED part of the conversation. Get a number of the parts and pieces, perhaps a dozen credits, and then launch LEED-EBOM. By then, you have moved people in the right direction. The key is not focusing on the [LEED] plaque, but rather on the true changes within the building and the environmental benefits behind those changes.

The Project Team

Creating and staffing the project team may be the first challenge. Many organizations hire an outside consultant with successful certification experience to help them organize the process for their first LEED-EBOM project. The team leader should have knowledge of the organization and credibility within the company. Hiring a new person to run a sustainability program and then giving them the LEED project as a first assignment is likely not going to work, even if that person has prior LEED project experience. The outside consultant can be used to facilitate the team meetings (Fig. 10.2), to keep the project on track, and to suggest ways to meet the various LEED requirements in the most efficient manner.

Sometimes, when the building is managed by a third party, that property manager can act as the LEED consultant and project lead on behalf of the client. Many of the

*Interview with Al Skodowski, March 2009.

Figure 10.2 **An outside consultant facilitates many team meetings, preferably someone with detailed knowledge not only of LEED credit requirements but also the documentation and certification process.** *Courtesy of Envision Realty Services.*

larger property management companies are setting up in-house LEED-EB teams as a client service. This is what happened when McDonald's set out to achieve LEED Platinum certification for one of the four buildings on its corporate campus in Oak Brook, Illinois (Fig. 10.3). While the on-site facility management team had a large role in the execution of the project, an outside LEED consultant, Helee Hillman of Jones Lang LaSalle (JLL) was hired to oversee the process from a budget, schedule, LEED administration and consulting standpoint, and acted as the lead for the project:*

> I recommend making sure that the client understands the expectation of their involvement upfront. Luckily on this project, we had a client that was really involved, very hands-on and really wanted to be a part of the effort every step of the way. But some other clients who have hired me to do projects think that since they signed the PO [purchase order] and hired you, they'll see you in six months with a [LEED] plaque [in your hands]. It's just not like that, because you need them to be engaged; because at the end of the day when the LEED consultant leaves the project, it's their job to make sure that all of the green practices that you've implemented are going to be

*Interview with Helee Hillman, May 2009.

Figure 10.3 In achieving a LEED-EB v.2.0 Platinum certification at its headquarters office building in 2009, McDonald's was motivated by its overall corporate sustainability commitment. *© McDonald's Corporation, photo by Ray Reiss.*

enforced and on-going to really maintain the spirit of getting LEED certification in the first place.

In other words, the LEED consultant is a facilitator and perhaps a subject-matter expert, but it's the organization's responsibility to keep up the green practices and operations. In the case of McDonald's, Hillman says it's important for the outside LEED consultant to engage and leverage the company's in-house resources:

> In our original budget, we thought we'd have to hire a handful of outside consultants around lighting, energy analyses, survey administration, data collection, air quality testing and so on, but we were able to leverage staff from the JLL in-house facility management team, the JLL in-house project managers as well as members from the McDonald's Facilities & Systems team. There were certain employees within the McDonald's team who wanted to help or wanted to get involved and ended up being a huge help to the process. It ended up being a team of 20 to 25 people that touched this project in one way or another which enabled us to eliminate the use of external consultants and save money on our bottom line because we were leveraging internal resources.

STAFFING THE PROJECT TEAM

What kind of expertise are you going to need on the LEED-EB project team? The answer: maybe 20 to 25 people, representatives of all affected departments, and particularly the people who will be doing the work. Make sure that all required expertise is included on the team; this may require participation from people in other locations, if you have centralized purchasing, legal, leasing, contracts, etc., as do many organizations.

Consultant Paul Goldsmith says that the team composition should be comprehensive:*

> One of the key aspects of a LEED project is putting the team together and making sure everyone knows their responsibilities. That includes everyone from the in-house team, the building engineer, V.P. of planning and facilities, and facilities managers to custodial staff, the outside maintenance staff and even the IT person. The IT person knows how many computers are in the building and that's a piece of data that you need to know.

Here are some of the most frequently used personnel resources:

- Facility manager
- Financial manager
- Building engineering and/or maintenance
- Energy manager
- Purchasing
- Custodial services
- Site and grounds maintenance
- Building cleaning services
- Environmental health and safety
- Building occupant representative

This might be the core team, added to as necessary to handle other issues that arise, such as waste management and recycling opportunities. Consultant Barry Giles likes to hand over responsibility for the project and process as soon as possible. He says:†

> What we're trying to do is create a team approach to understanding where you are now and where you want to get to, because the last thing we want to do as a consultant is be the only person in the room doing this. We would never be able to stay in business long enough to do that because we would be too tied up with one job, so we've got to get them to understand that this is their project, not our project. It's their building.
>
> We devised a method of segmenting the whole LEED-EB process and dividing it into bite-sized chunks. We have a seven- to eight-phased process for doing the certification. That gives the client control. It gives them a swing gate at the end of every phase, where they can [continue with the process or] say, "I like what's happened, but we don't have the finances to go any further. Let's stop right now."

---

*Interview with Paul Goldsmith, March 2009.

†Inteview with Barry Giles, April 2009.

The Building Audit—Getting the Information Together

Sometimes the first major project challenge is just finding all the information you're going to need to demonstrate compliance with various LEED-EBOM credits. Consultant Elaine Aye of Green Building Services knows this situation well.*

> One of the first challenges is in understanding how the base building performs. A lot of building owners outsource facility management services. It may be landscaping maintenance services, mechanical system upgrades, preventive maintenance and so forth. One of the biggest problems is having a common database of building information in one location. So when we certify a building, we find out how the building performs, the utility bills, the water bills, etc. All of that may not be in a common location. We also want to know the kind of light bulbs purchased in the last 12 months. Well, they may have 20 different kinds of light bulbs in their building and no record of the specifications. Having a common database is an important aspect of the process. Having someone [on the project team] who is really knowledgeable about the building consistently is an important aspect [of the certification process], and that doesn't always happen.

The building audit is a critical part of the process. It's the first chance to compare your accomplishments with the LEED requirements. You can look at this stage for no-cost/low-cost measures that will help garner perhaps 10 to 12 LEED-EBOM points, and you can use the LEED scorecard as a checklist for tallying these points and for managing the process.

MEETING THE LEED PREREQUISITES

Of course, the first thing the team leader and consultant should do is to determine that the project can meet all of the LEED-EBOM prerequisites. As we've indicated, the minimum energy performance and minimum ventilation performance are probably the two most difficult prerequisites to meet, especially in older buildings. Four of the others require writing and adopting policies, which is not difficult for most organizations. Consultant Barry Giles talks about the LEED prerequisites as "showstoppers":

> There are three important showstoppers. The most important is EQ prerequisite 1, which mandates testing fresh air delivery into the building. That poses significant financial difficulties for some projects. In LEED-EBOM there is an air-handling [minimum ventilation] requirement that is causing financial difficulty because it requires every single air handler to be tested. In some buildings, that can mean up to 170 air handlers need to be tested. That costs money and, as a prerequisite, it has to be done. Clients are balking about it a bit, especially in this recession.

---

*Interview with Elaine Aye, March 2009.

> It's fine in class A buildings where you get an ENERGY STAR rating. In a class A building—75 percent of the time you can get a passing mark. But as we move down the grade into class Bs and class Cs, getting that ENERGY STAR rating is hard and is the forerunner for people saying, "I can't do LEED because I cannot get an ENERGY STAR rating."

LEED-EBOM works best now for well-constructed "Class A" office buildings, dating from the 1990s and 2000s. (This may mean, unfortunately, that older office buildings may not see any energy upgrades or green certifications for some time.) However, some building owners such as the federal GSA, some states and cities, some universities and large developers have older buildings that were built to these contemporary standards, or have been continuously well maintained and periodically upgraded to new standards and technology. As illustrated in the case studies presented in Chap. 13, these types of buildings can also pass these three "showstopper" tests without significant capital investments.

Organizing the Project

Grouping the various LEED credits into functional categories makes it easier to divide responsibilities among team members.* For example, the Materials and Resources (MR) credits 1 through 5 all require purchasing policies and documentation of purchases during the performance period, so the head of purchasing is the logical person to take on that task. In the same way, MR credits 6 through 9 all involve the "back end" of the operation, waste management policies and implementation actions, which should logically be the province of the building manager or facility manager. Other key functional areas include:

- Administration, planning, and logistics, which will deal with occupant surveys, commuting arrangements, space layout for daylight and views, and assessing cost and financial implications.
- The indoor environmental quality (EQ) green cleaning credits may affect janitorial contracts, product purchases, and equipment purchases, and therefore, they need to be coordinated with contracts/purchasing departments.
- Site management credits may be the province of a facility or building manager. In many cases, such as Owens Corning's headquarters in Toledo, Ohio, site management has been contracted to an outside organization. In such cases these contractors need to be brought on board with the project team.
- Occupant health and safety issues in the EQ section may be the joint province of an environmental health and safety (EHS) officer as well as a facility manager.

*LEED Reference Guide, p. xvii.

Case Study: SMUD Customer Service Center, Sacramento, California

At the Sacramento Municipal Utility District (SMUD), architect Brian Sehnert was the in-house champion for taking a very energy-efficient building and achieving a LEED-EB Platinum certified project in 2006 and 2007.* Designed by Williams + Paddon Architects + Planners Inc., the SMUD's Customer Service Center was built with green features in 1995 (Fig. 10.4), well before LEED existed, with additional sustainable features and maintenance practices incorporated in 2006. Examples of the facility's improvements include:

- Installing devices to better control indoor humidity and temperature levels
- Fitting carbon dioxide sensors into conference rooms, training rooms, and other high-density areas to reduce air-conditioning needs when rooms are unoccupied
- Installing water meters and collecting data to eliminate unnecessary irrigation

Figure 10.4 **SMUD's Customer Service Center was built in the mid-1990s with many green features and upgraded in 2007 to LEED-EB Platinum status.**
Courtesy of Wililams + Paddon Architects + Planners Inc.

*www.smud.org/en/news/pages/news07.aspx#jul, SMUD Press Release, August 3, 2007, accessed May 31, 2009. Also interview with Doug Norwood, April 2009.

- Installing energy meters in each wing of the Service Center to pinpoint areas of high-energy usage for potential energy reduction projects
- Adoption of new strategies to maximize the building's daylighting features such as exterior light shelves, which are pressure-washed twice a year to maintain high reflectivity

After the certification effort at the Customer Service Center, SMUD adopted new sustainability maintenance practices that will be extended to all of the utility's buildings and facilities throughout Sacramento County. These included:

- Switching to environmentally friendly cleaning products, integrated pest management, and non-chemical fertilizer solutions
- Stocking restrooms with recycled toilet tissue and paper towels
- Mulching all green waste on-site, which is one of the required LEED green site maintenance practices
- Using only recyclable and low-VOC (volatile organic compound) construction materials, including carpet and paint, for additions, renovations, and remodels

Sometimes, it is not easy meeting the LEED-EBOM credit requirements because documentation is just hard to come by. Even for credits that on the surface should be easily obtained, because the underlying practices meet LEED standards, quite a bit of documentation may have to be developed from scratch on many projects. In case of the SMUD project, according to Doug Norwood:

> We struggled a little bit with our green exterior site management, for instance with the plants used in the landscaping. Even though when the building was built, they put in as much native and adapted plants as possible, documenting that was very difficult. Not only going around and documenting every plant, but also finding an approved list of native and adaptive plants for our area was a real struggle. It took us a long time to find that. You have to somehow be able to show that the plantings around your building are native to the area.

The Gap Analysis

The building audit ought to reveal the "gap" between how the building or facility is performing at the present time and where you want it to be in the future, to meet the LEED requirements. The gap analysis typically follows the building audit and helps teams to apply some costs to the various measures they'll have to adopt. It functions as a "LEED-EB feasibility study" that may be needed to get upper management to provide the staff time and funds for the certification effort. It also helps the team to see the process in a more holistic way. Green Building Services' Elaine Aye comments:

> Another challenge [in many projects] is that people aren't talking. Everybody has a piece of the pie in a building, but there's no central direction. Everyone focuses on their own area and not the entire process of the building. One of the opportunities that we bring to

Figure 10.5 **Interface, Inc.'s 280,000-square-foot Bentley Prince Street carpet factory in Los Angeles received a LEED-EB Silver certification in 2007.*** **Project highlights include all electricity produced 100 percent by renewable resources, including a 100-kilowatt solar array and Green-e certified renewable energy purchases; a comprehensive sustainability education program for employees, customers, suppliers, and community partners; an on-site waste diversion program, with 95 percent recycling of all waste; and 95 percent of purchases meeting sustainability criteria.** *Courtesy of Bentley Prince Street.*

the table is helping the group communicate differently. Create a green team from your tenants, your employees and the people that service your building to work together to create a common plan and a goal to help address your building's efficiencies or inefficiencies. Then set benchmarks for how you want the building to perform in the future.

Managing the Certification Process

Once you've made the decision to go forward with the project, getting started with the actual certification process requires leadership, because of the implied commitment to an end result. Registering the project with the Green Building Certification Institute (a 2008 offshoot of the USGBC that oversees the LEED certification process) is a bit like getting engaged: you have to tell your friends and family of your intentions and

*Case study provided to the author by Kimbrely Matsoukas, Bentley Prince Street.

go public with the relationship and implied commitment to the end result. But it's a necessary first step toward accessing all of the online templates that the project team needs to access to start providing the required documentation.

THE PERFORMANCE PERIOD

Now is the time to begin holding regular project meetings and committing to a performance period. Many of the consultants interviewed felt that most projects could finish inside of nine months from the decision to proceed. The real key is to decide on the certification level you want to achieve, prepare a realistic budget, figure out a way to track costs, and assign clear responsibilities to each member of the project team. This might be a good time to educate team members about the LEED documentation requirements, the entire certification process, and how to get rulings on difficult credits.

The retro-Cx is clearly an early stage requirement, because its results will guide much of the energy and ventilation improvement projects. Utility bills for the past year need to be compiled and made available to the project team. The retro-Cx combines with baseline information such as an ENERGY STAR score and water usage, to create a "gap analysis" between where the project is now and where it needs to be. An early consideration is to put purchasing, waste management, and green cleaning policies into place right away and begin to measure activity in those areas. Decide which quarter (the minimum performance period) of the year will represent the performance period or time of implementation. While tempting on some projects, it's not wise to start the clock running until you've done whatever upgrades will be required for achieving certification. Elaine Aye says:

> If you're going after certification, you obviously have to meet the system's prerequisites first, and once you've done that, you can push the envelope to see what other areas can be addressed. Once we establish that, we go through an implementation period. The implementation period is: You've upgraded the building, made capital investments, upgraded the chillers, upgraded the plumbing, upgraded the lighting systems, implemented your strategies and now you're going to track and record the performance based on those upgrades. You're going to implement your polices and do the necessary training and education for staff and vendors. You're going to involve tenants and employees and pool all of the parties in the building together to start tracking and recording how the building is doing.

Managing the Process from Beginning to End

Consultant Paul Goldsmith has managed a number of LEED projects. He says the key is getting everyone to work together, but it's important to observe some of LEED-EBOM's peculiarities, such as the performance period:*

*Interview with Paul Goldsmith, March 2009.

> Bringing the team together is key and then sitting down and doing, for all intents and purposes, a checklist review which we'll call a charrette. Everybody sits down and marches through the various credits and will weigh in on them whether they're achievable, partially achievable or definitely not something to pursue. From the charrette, you then get clear direction from the ownership or upper management and everybody knows what the direction is to achieve the goal.
>
> Whether you strive to just get [basic] Certified or you pick a [higher] level of certification that you're striving for, that's something that the team will take direction from the ownership. I always say, "Don't chase credits for credits' sake but do it for the right reasons and from an environmental stewardship point of view." They all must work together from an environmental stewardship, social awareness and economic [point of view].
>
> It's a three-legged stool. You can't take one leg away because it will fall over. You've got to have them all pulling at some equal level so there's a balance within the team and everybody is pulling in the same direction. So getting that charrette done is key. Getting the checklist done is key. Getting registered early and then delegating the responsibilities, meeting the timelines and then breaking up those credits into the three basic categories [environmental, economic and social] is key. Once all of those are in play, you start working on them and then the LEED Online system is used as a project management tool to march the team step-by-step down the road. Then you'll set some timelines to get to a target that you can all work toward, to get the project ready for registration and certification. Once you get everything done, sometimes at the end, you're still dotting the i's and crossing the t's during the performance period.

The LEED scorecard is a fundamental tool used by all project teams for tracking progress toward certification. Figure 10.6 shows the final LEED scorecard for the FBI Regional Office in Chicago, developed and managed by Transwestern. The project achieved a LEED-EB version 2.0 Platinum certification in December 2008.

You can see from Paul Goldsmith's testimony that experience with the LEED-EB project management process is critical to a successful outcome. Consultant Craig Sheehy likes to manage his projects within a compressed, 20-week process that includes the performance period (13 weeks) and about a month on either side. Then he tries to do most of the work with his team, so that each client project team member is only committing to one or two hours a week to specific LEED-EB certification work.*

> We give them a timeline, we break it down during the 20 weeks and each week we're going to work on an item. For stuff that we need from them, we'll give examples from other buildings, so they're not starting from scratch. There's a three-month performance period that we take these buildings through. We start 30 days prior to the beginning of the performance period collecting data on the building—everything from site plans, floor plans, building information, having the engineer do a sequence of operations, preventive maintenance plans, listing of the equipment and the number of fixtures and lights.

*Interview with Craig Sheehy, March 2009.

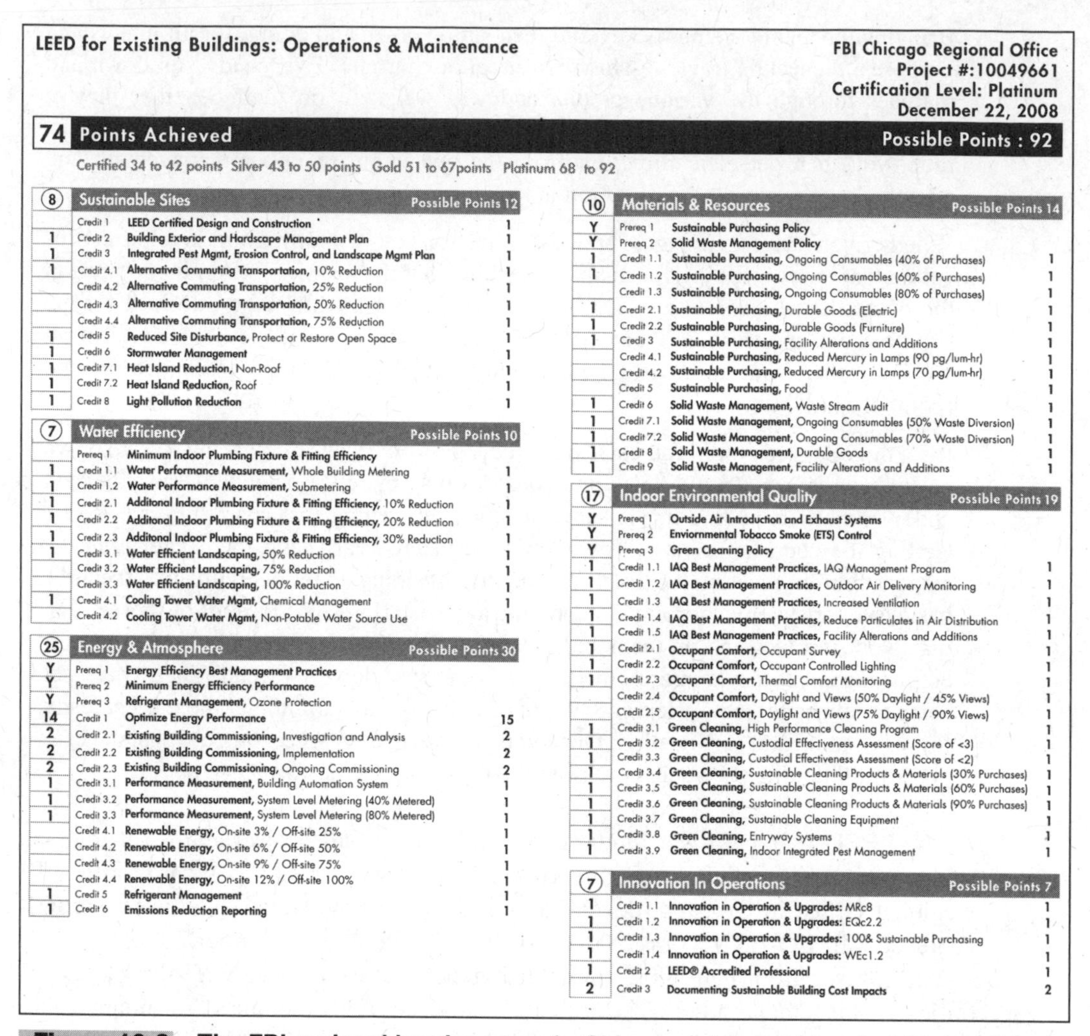

LEED for Existing Buildings: Operations & Maintenance

FBI Chicago Regional Office
Project #:10049661
Certification Level: Platinum
December 22, 2008

74 Points Achieved — Possible Points : 92

Certified 34 to 42 points Silver 43 to 50 points Gold 51 to 67points Platinum 68 to 92

8	Sustainable Sites		Possible Points 12
	Credit 1	**LEED Certified Design and Construction**	1
1	Credit 2	**Building Exterior and Hardscape Management Plan**	1
1	Credit 3	**Integrated Pest Mgmt, Erosion Control, and Landscape Mgmt Plan**	1
1	Credit 4.1	**Alternative Commuting Transportation,** 10% Reduction	1
	Credit 4.2	**Alternative Commuting Transportation,** 25% Reduction	1
	Credit 4.3	**Alternative Commuting Transportation,** 50% Reduction	1
	Credit 4.4	**Alternative Commuting Transportation,** 75% Reduction	1
1	Credit 5	**Reduced Site Disturbance,** Protect or Restore Open Space	1
1	Credit 6	**Stormwater Management**	1
1	Credit 7.1	**Heat Island Reduction,** Non-Roof	1
1	Credit 7.2	**Heat Island Reduction,** Roof	1
1	Credit 8	**Light Pollution Reduction**	1

7	Water Efficiency		Possible Points 10
	Prereq 1	**Minimum Indoor Plumbing Fixture & Fitting Efficiency**	
1	Credit 1.1	**Water Performance Measurement,** Whole Building Metering	1
1	Credit 1.2	**Water Performance Measurement,** Submetering	1
1	Credit 2.1	**Additonal Indoor Plumbing Fixture & Fitting Efficiency,** 10% Reduction	1
1	Credit 2.2	**Additonal Indoor Plumbing Fixture & Fitting Efficiency,** 20% Reduction	1
1	Credit 2.3	**Additonal Indoor Plumbing Fixture & Fitting Efficiency,** 30% Reduction	1
1	Credit 3.1	**Water Efficient Landscaping,** 50% Reduction	1
	Credit 3.2	**Water Efficient Landscaping,** 75% Reduction	1
	Credit 3.3	**Water Efficient Landscaping,** 100% Reduction	1
1	Credit 4.1	**Cooling Tower Water Mgmt,** Chemical Management	1
	Credit 4.2	**Cooling Tower Water Mgmt,** Non-Potable Water Source Use	1

25	Energy & Atmosphere		Possible Points 30
Y	Prereq 1	**Energy Efficiency Best Management Practices**	
Y	Prereq 2	**Minimum Energy Efficiency Performance**	
Y	Prereq 3	**Refrigerant Management,** Ozone Protection	
14	Credit 1	**Optimize Energy Performance**	15
2	Credit 2.1	**Existing Building Commissioning,** Investigation and Analysis	2
2	Credit 2.2	**Existing Building Commissioning,** Implementation	2
2	Credit 2.3	**Existing Building Commissioning,** Ongoing Commissioning	2
1	Credit 3.1	**Performance Measurement,** Building Automation System	1
1	Credit 3.2	**Performance Measurement,** System Level Metering (40% Metered)	1
1	Credit 3.3	**Performance Measurement,** System Level Metering (80% Metered)	1
	Credit 4.1	**Renewable Energy,** On-site 3% / Off-site 25%	1
	Credit 4.2	**Renewable Energy,** On-site 6% / Off-site 50%	1
	Credit 4.3	**Renewable Energy,** On-site 9% / Off-site 75%	1
	Credit 4.4	**Renewable Energy,** On-site 12% / Off-site 100%	1
1	Credit 5	**Refrigerant Management**	1
1	Credit 6	**Emissions Reduction Reporting**	1

10	Materials & Resources		Possible Points 14
Y	Prereq 1	**Sustainable Purchasing Policy**	
Y	Prereq 2	**Solid Waste Management Policy**	
1	Credit 1.1	**Sustainable Purchasing,** Ongoing Consumables (40% of Purchases)	1
1	Credit 1.2	**Sustainable Purchasing,** Ongoing Consumables (60% of Purchases)	1
	Credit 1.3	**Sustainable Purchasing,** Ongoing Consumables (80% of Purchases)	1
1	Credit 2.1	**Sustainable Purchasing,** Durable Goods (Electric)	1
1	Credit 2.2	**Sustainable Purchasing,** Durable Goods (Furniture)	1
1	Credit 3	**Sustainable Purchasing,** Facility Alterations and Additions	1
	Credit 4.1	**Sustainable Purchasing,** Reduced Mercury in Lamps (90 pg/lum-hr)	1
	Credit 4.2	**Sustainable Purchasing,** Reduced Mercury in Lamps (70 pg/lum-hr)	1
	Credit 5	**Sustainable Purchasing,** Food	1
1	Credit 6	**Solid Waste Management,** Waste Stream Audit	1
1	Credit 7.1	**Solid Waste Management,** Ongoing Consumables (50% Waste Diversion)	1
1	Credit 7.2	**Solid Waste Management,** Ongoing Consumables (70% Waste Diversion)	1
1	Credit 8	**Solid Waste Management,** Durable Goods	1
1	Credit 9	**Solid Waste Management,** Facility Alterations and Additions	1

17	Indoor Environmental Quality		Possible Points 19
Y	Prereq 1	**Outside Air Introduction and Exhaust Systems**	
Y	Prereq 2	**Enviornmental Tobacco Smoke (ETS) Control**	
Y	Prereq 3	**Green Cleaning Policy**	
1	Credit 1.1	**IAQ Best Management Practices,** IAQ Management Program	1
1	Credit 1.2	**IAQ Best Management Practices,** Outdoor Air Delivery Monitoring	1
1	Credit 1.3	**IAQ Best Management Practices,** Increased Ventilation	1
1	Credit 1.4	**IAQ Best Management Practices,** Reduce Particulates in Air Distribution	1
1	Credit 1.5	**IAQ Best Management Practices,** Facility Alterations and Additions	1
1	Credit 2.1	**Occupant Comfort,** Occupant Survey	1
1	Credit 2.2	**Occupant Comfort,** Occupant Controlled Lighting	1
1	Credit 2.3	**Occupant Comfort,** Thermal Comfort Monitoring	1
	Credit 2.4	**Occupant Comfort,** Daylight and Views (50% Daylight / 45% Views)	1
	Credit 2.5	**Occupant Comfort,** Daylight and Views (75% Daylight / 90% Views)	1
1	Credit 3.1	**Green Cleaning,** High Performance Cleaning Program	1
1	Credit 3.2	**Green Cleaning,** Custodial Effectiveness Assessment (Score of <3)	1
1	Credit 3.3	**Green Cleaning,** Custodial Effectiveness Assessment (Score of <2)	1
1	Credit 3.4	**Green Cleaning,** Sustainable Cleaning Products & Materials (30% Purchases)	1
1	Credit 3.5	**Green Cleaning,** Sustainable Cleaning Products & Materials (60% Purchases)	1
1	Credit 3.6	**Green Cleaning,** Sustainable Cleaning Products & Materials (90% Purchases)	1
1	Credit 3.7	**Green Cleaning,** Sustainable Cleaning Equipment	1
1	Credit 3.8	**Green Cleaning,** Entryway Systems	1
1	Credit 3.9	**Green Cleaning,** Indoor Integrated Pest Management	1

7	Innovation In Operations		Possible Points 7
1	Credit 1.1	**Innovation in Operation & Upgrades:** MRc8	1
1	Credit 1.2	**Innovation in Operation & Upgrades:** EQc2.2	1
1	Credit 1.3	**Innovation in Operation & Upgrades:** 100& Sustainable Purchasing	1
1	Credit 1.4	**Innovation in Operation & Upgrades:** WEc1.2	1
1	Credit 2	**LEED® Accredited Professional**	1
2	Credit 3	**Documenting Sustainable Building Cost Impacts**	2

Figure 10.6 The FBI regional headquarters in Chicago achieved 14 out of 15 possible points for energy efficiency en route to a LEED Platinum certification.

Then the performance period starts. During the performance period, it's pretty laid back for them. They're just sending us invoices [for utility bills and building purchases and services] as they go. We'll be working with the janitorial company, the purchasing group (lamps and chemicals and so forth). Then once the performance period ends, we get an easy point by just showing operating expenses of the building. We need five years prior operating expenses and the operating expenses during the performance period. So if the performance period ends in June 2009, those financials are going to come around the 20-25th of July. That's normally the last piece. As soon as those financials come out, they upload them to us and we hit the submit button [and wait for the review].

Project Profile: Owens Corning World Headquarters, Toledo, Ohio

Originally built in 1995, Owens Corning's headquarters was certified LEED-EB version 2.0 Silver in 2006 (Fig. 10.7).* When Owens Corning decided to apply for LEED-EB certification, the company sought help from Johnson Controls, Inc. (JCI), which developed a return on investment planner for the facility using JCI's Green Compass™ building assessment and management tool. JCI's initial look at the building found 23 LEED credits that were either already earned or achievable with documentation. The team found the means to earn 21 more credits for total savings of more than $100,000 per year. Owens Corning implemented and documented the additional energy saving measures, and then engaged Johnson Controls to compile the LEED-EB submittal. The facility is maintained and operated by Hines Management Services.

Figure 10.7 Owens Corning's World Headquarters in Ohio received LEED version 2.0 Silver certification in 2006, with more than 50 percent of the site restored to open space. More than 90 percent of occupants have views to the outdoors from their workstations, and an underfloor air distribution system provides individual temperature controls. *Courtesy of Owens Corning.*

*Information from Owens Corning, "Pink Green Silver: Owens Corning World Headquarters Certified for Sustainability," Publication No. 10005543, October 2007.

With help from Hines, Owens Corning has reduced overall electrical use every year since the building was commissioned except one, and that was when a new and more-powerful computer system was first installed.

Innovation in Operations

Throughout the discussion of various LEED credits, we've referenced additional credits available for "exemplary performance," that is, going above and beyond specific LEED credit requirements. LEED-EBOM 2009 reserves 10 points for actions that promote "innovation in operations." The USGBC regards these innovation credits as an important way for the LEED rating system to "keep pace with the evolving practices and standards of the industry."* These points can often make the difference between one level of certification and the next higher level, and teams typically focus on them starting at the midpoint of most projects. There are four components to this section of the LEED-EBOM rating system, as shown in Table 10.1.

EXEMPLARY PERFORMANCE AND INNOVATIONS

Potential credit points can be earned for "exemplary performance" in existing LEED categories; for example, achieving an ENERGY STAR score of 97 would garner one point in this category.† Not all LEED credits are eligible for exemplary performance, so project teams must consult the LEED Reference Guide to determine for which of the potential "exemplary performance" activities they plan to seek LEED points, if

TABLE 10.1 INNOVATION IN OPERATIONS

CREDIT CATEGORY	DESCRIPTION
1. Innovation in operations	a. Exemplary performance: 1 to 3 points
	b. Achievements not addressed by LEED credits: 1 to 4 points
	c. 4 points maximum in this credit category
2. LEED Accredited Professional (LEED AP)	1 point for having a principal participant of the project team with LEED AP status
3. Documenting sustainable building cost impacts	1 point for documenting changes in building operating costs during the performance period
4. Regional priority	1 to 4 points for credits in 6 targeted regional priority categories

*LEED Reference Guide, p. 465.

†Ibid., p. 164.

any. What is exemplary performance? This term means achieving all of the LEED points in a given category and then going beyond the top tier to the next logical step. For example, if one gains 5 water conservation points for achieving 30 percent reduction in water use, then one can garner an additional LEED point with a documented 35 percent reduction.*

In addition, LEED awards up to 4 points for doing things considered worthwhile in LEED terms, but are not covered by the current credits. This may include things that are valuable such as public education about the LEED certification or even "borrowing" credits from other LEED 2009 systems. The USGBC's guidance on this issue is that strategies and measures must have significant environmental benefits equivalent to other LEED credits; the project team must define, execute, and document the initiative; the strategy must be comprehensive and effective at achieving the stated environmental benefit; and strategies should integrate the building's operations across technical categories. In addition, the innovative concept "must rise significantly above standard building operations and maintenance practices."†

A project team can combine these two approaches to earn up to 4 points in this section, pick up perhaps 2 exemplary performance credit points and 2 innovation credit points. This is an area where an experienced LEED consultant can be especially useful in helping the project team to choose from among many possibilities.

LEED ACCREDITED PROFESSIONALS

To be eligible for this point, the LEED AP must have specifically passed the LEED-EBOM national exam, administered by the Green Building Certification Institute‡ and must be a "principal participant" in the project team.

DOCUMENTING SUSTAINABLE BUILDING COST IMPACTS

The USGBC has a strong interest in building a business case for the LEED-EBOM system and so it offers 1 credit point for project teams that provide data on the changes in operating costs from the measures adopted in the certification process and the financial impacts of various investments made to achieve operating cost reductions. The credit requires teams to document overall building operating costs for the previous five years (or length of occupancy if shorter). The LEED system gives this credit because the USGBC is particularly interested in documenting savings in the water efficiency and energy and atmosphere categories, resulting from measures taken to achieve compliance with particular credit requirements.§

---

*Not all LEED credits qualify for exemplary performance, so project teams should consult the LEED Reference Guide to strategize which to pursue.

†LEED Reference Guide, p. 469.

‡The Green Building Certification Institute has taken over the testing and certification tasks under the LEED Rating System from the USGBC, but remains closely allied with it. www.gbci.org, accessed May 31, 2009.

§LEED Reference Guide, p. 478.

MEETING REGIONAL PRIORITIES (RP)

All of the LEED rating systems represent green templates suitable for all building types and sizes, all ownership types, and all regions of the United States. As a result, LEED has often been criticized for failing to give more consideration to regionally significant environmental issues, for example, water in the Southwest. One of the more innovative changes in the LEED v3 (2009) system is the provision of 4 points for meeting regional priorities (identified by the USGBC chapters). For each project zip code, the regional priority credits identify 6 of the existing LEED-EBOM credits that, if achieved, will automatically garner an extra point from this category. Up to 4 points may be achieved in this way. Again, an experienced LEED consultant can help the project team focus on which of the RP credits they want to pursue. The USGBC Web site contains a searchable database of these credits.* At this time, projects outside the United States are not eligible for these credits.

Summary

LEED certification at any level is a significant attainment and from all reports, well worth the effort. The keys to a successful project are creating a well-integrated project team, implementing a process that gets the job done in the minimum time possible, and having a realistic budget, based on the building's or facility's current situation. Some LEED-EB projects have been virtually free, except for staff time, while others have cost up to $500,000. In the next chapter, we take the lessons learned from successful certification projects and outline 10 "best practices" that will help your future LEED-EBOM project proceed smoothly and generate cost-positive results.

*http://www.usgbc.org/DisplayPage.aspx?CMSPageID=1984, accessed May 31, 2009.

11

LESSONS LEARNED—TEN BEST PRACTICES FOR GREENING EXISTING BUILDINGS

While writing this book, I came across a remarkable process at General Electric called the Treasure Hunt, and it got me thinking that LEED-EB is like a treasure hunt for environmental sustainability. At GE, the objective is to cut operating costs and carbon by having a team of GE staff with expertise in areas such as facility management, energy, lean manufacturing, finance, etc., combined with external experts, dive into an operation at some facility and, in the course of a few days, working with the local facility staff, find real, measurable, long-term, and cost-effective savings. The Treasure Hunt is an adaptation of the Toyota Energy Hunt process from lean manufacturing and a clear illustration of what LEED-EB is meant to create: *Kaizen*, or continuous improvement, in this case applied to building energy and water use.

GE's Treasure Hunt

Gretchen Hancock is a project manager for corporate environmental programs at GE, and she runs the Treasure Hunt process. I think her observations provide a good mental model for every LEED-EB process and one worth emulating by project teams.*

> One of our sayings is "We'll play with whoever wants to play with us." So you get people who are energized about the concepts of energy efficiency, greenhouse gas reduction and [GE's] *ecomagination*^{sm} [initiative] in general who want to come learn about the process.

*Interview with Gretchen Hancock, May 2009.

> The best treasure hunts happen with incredibly diverse teams of people. It's key that you have the people in the facility who have what I call, "the biggest key rings." They're the maintenance people that have keys to all of the access rooms, the roof and those kinds of places. You need those people. It's really good to have process-thinkers on the same team—people who know the manufacturing processes, who might have a background in disciplines like lean manufacturing or Six Sigma to be able to augment the actual mechanical expertise that the other folks bring.

Here Hancock emphasizes one of the key points made by most LEED-EB consultants and project teams: the importance of having a diverse set of eyes, diverse skill sets, and a good mix of practical and theoretical knowledge.

> We've also found that it's great to have people participate in a treasure hunt in a location that they're not associated with, because those are the people that will ask the "why?" questions: Why is this equipment operating the way it is? They can really challenge the status quo so the more diverse the team is, the better.

In a LEED-EB project, it's vital to have the retro-Cx done by someone who may be familiar with the building type and HVAC systems it uses, but who is new to the facility, so they can ask similar "WHY?" questions.

> One of the critical pieces to a treasure hunt is getting buy-in from the equipment operators, and the process is designed to do that. What we don't do at the end of the treasure hunt is have a list of projects that look great from a facilities perspective but are completely unworkable from a manufacturing perspective. So a key part of the treasure hunt process is getting feedback and engagement from those the manufacturing team leaders.

In a similar vein, a LEED-EB project that does not engage building operations and maintenance staff in supporting the recommended, planned, and budgeted improvements is likely to fail in the long run. Hancock says some projects get an immediate "go ahead," while others need to wait for budgets to become available:

> One of the things we're getting our hands around now is what has been implemented. Generally, sites will implement about one-third of the projects because when we look across the data set as a whole, about a third of the projects will fall into a category that site leadership teams call "Just Do It" projects—those that payback within less than nine months.

In the same way, we're finding that many LEED-EB projects lead immediately to no-cost/low-cost upgrades that can be completed quickly and paid for within an annual maintenance budget. Sometimes, of course, a treasure hunt can uncover some real buried treasure. Hancock relates one incident that really stands out:

> One of the interesting things that happened recently at the Waukesha [Wisconsin] facility was when we found one single project that would reduce water usage at the

facility by 45 percent. It involved a fairly big expenditure and had a three-year payback, but the site leadership team in concert with the finance team said, "This project makes a lot of sense. It's the right thing to do, so let's put that on a long-term schedule for implementation and break it down into smaller projects to work on it over the next few years and go from there." While there's no requirement, we see a lot of projects being implemented both for cost-savings reasons and for reasons where people are simply making the right choice around energy efficiency and resource utilization.

Even though they are focused mainly on factory operations, some of the treasure hunts have similar outcomes as building energy audits, in terms of what needs to be changed.

One common change that we identify as a result of treasure hunts is relamping. That's a very typical project. Also, installing variable-frequency drives (VFDs) in both production equipment and HVAC equipment pays back considerably in terms of energy efficiency. Compressed air leakage is the one that continues to surprise us, as to the value to which you are making sure your compressed air lines in a facility don't leak. Those projects often have a great payback so we'll often identify incremental changes that can be made, which from an aggregate standpoint can make a big difference around energy efficiency.

Finally, Hancock believes that the treasure hunts can (and do) serve other purposes, in terms of getting people to look at their facility differently in the future.

The treasure hunt is more than anything about culture change, and it results in a great repository of projects. It also really energizes the site teams and operation to think about energy utilization in a different way. We have done several of what we call "Treasure Hunt 2s" at sites that have already been through an initial treasure hunt process, and we found an additional 20 percent opportunity to reduce energy expenditures. I think that speaks to the strength of the process. It also shows how much GE operations grow and change over a three- or four-year period.

According to Hancock:

GE has trained more than 3500 of its employees globally to think about wasted energy and water in a different and powerful way. Those individuals have identified more than 5000 projects that have the opportunity to drive energy efficiency, eliminate 700,000 metric tons of greenhouse gas emissions—and $111 million in operational cost.*

*www.greenerbuildings.com/blog/2009/05/13/ge-treasure-hunts-discover-millions-in-savings, accessed May 31, 2009.

Hancock's "lessons learned" are summarized in two succinct observations:*

1. In addition to operator-buy in, there are two more critical elements of staffing an energy treasure hunt. First, facility leadership must be committed to implementing some of the projects identified during the event.
2. Second, the most successful events leverage internal operational, environmental, health, and safety and maintenance expertise side by side with representatives of utilities, contractors, and people from other locations or companies—experience combined with fresh eyes on the process results in the right questions being asked.

LEED-EBOM's requirement for recertification within a one- to five-year window after the initial certification† (as called for in LEED v3) is designed to help each building or facility stay current with new thinking about environmental attributes, with new technologies, and with current economic conditions. In this way, LEED promotes a continual "treasure hunt" among its certified projects.

THE FEDERAL APPROACH TO TREASURE HUNTS

The U.S. General Services Administration (GSA) reported in 2009 on seven strategies that promote cost-effective energy savings and productivity gains in federal buildings, shown in Table 11.1. In this way, GSA has published the results of its own treasure hunt. With these simple strategies, fully implemented in all federal buildings, GSA can save $57 million in annual operating costs. The GSA Workplace Performance Study surveyed over 6000 federal workers and measured environmental conditions at 624 workstations in 43 workplaces in 22 separate buildings.‡ More than 60 percent of survey respondents were unhappy with workplace temperatures, suggesting lower productivity. In fact, GSA's study found wide variations in temperatures from the prescribed settings.§ Better air filtration also led to higher satisfaction, indicating people are quite sensitive to dust in the work environment, even to particles as small as 2.5 microns. Since worker costs (salaries and benefits) dwarf energy costs more than 100 to 1 in most knowledge organizations ($300 to $600/year/square foot vs. $3.00/year/square foot), it's logical to assume that implementing the treasure hunt will have a large benefit, not only on energy use, but also on the much larger issues of increasing productivity and health.

You can see that some of these measures are easy to implement and could fall within the purview of a LEED-EB certification effort, such as upgrading ambient and task lighting, adjusting summer office temperatures, and replacing HVAC filters, while others that are quite effective, such as consolidating printers and copiers and replacing all older-style monitors and computers, may have to await better financial times. Still other

---

*Ibid.

†LEED Reference Guide, p. xvii.

‡*Energy Savings and Performance Gains in GSA Buildings*, March 2009, Public Buildings Service, General Services Administration. Savings are based on 176 million square feet of space and average electricity cost of $0.10 per kilowatthour. Available at www.wbdg.org/research/energyefficiency.php?a=11, accessed May 10, 2009.

§Ibid., p. 5.

TABLE 11.1 SEVEN STRATEGIES FOR ENERGY SAVINGS AND PERFORMANCE GAINS IN GSA BUILDINGS

STRATEGY	RELATIVE COST	ENERGY SAVINGS (kWh/YEAR)	ENERGY COST SAVINGS ($/YEAR)
1. Adjust summer workplace temperatures	$ (low)	18.7 million	$1.87 million
2. Replace HVAC filters on schedule and with high performance filters	$	10.8 million	$1.08 million
3. Consolidate and reduce number of printers and copiers	$$ (medium)	55.0 million	$5.5 million
4. Replace CRT monitors with flat-panel LCD monitors	$$	39.0 million	$3.9 million
5. Upgrade ambient and task lighting	$$	199.1 million	$19.9 million
6. Improve access to daylight in workplace	$$$ (high)	118.1 million	$11.8 million
7. Upgrade windows for better energy performance	$$$	127.5 million	$12.8 million
Total all measures		**568.2 million**	**$56.82 million**

measures, such as upgrading windows, will probably have to await major building renovations such as we saw with the example of the Empire State Building.

Don Horn is director of the Sustainability Program of GSA's Public Buildings Service. About integrating the clear mandate for federal energy-efficiency programs with LEED-EB, he says:*

> We have about 24 LEED-EB projects (total registered and certified) (out of 164 total LEED projects). We have three certified LEED-EB buildings—The John J. Duncan Federal Building and the Byron G. Rogers U.S. Courthouse, that are government owned; and the FBI building in Chicago which is leased and LEED-EB Platinum. It was interesting to note that it was the building ownership [USAA] that decided to pursue LEED-EB, so we didn't have a lot of input in that.
>
> We don't require LEED-EB right now. We have not made a decision to use LEED-EB for our entire inventory. We encourage it and we would like to see our individual buildings working towards a LEED-EB certification, but we haven't decided to do that on a wholesale basis. Our federal mandate is to follow a set of guiding principles for high-performance, sustainable buildings. Some of them align with the LEED-EB credits, but LEED-EB has far more requirements than the guiding principles, so it's beyond our basic mandate.

*Interview with Don Horn, March 2009.

The buildings that have pursued or are pursuing LEED-EB are the ones in which the building manager, project manager, or regional office decided that they wanted to get LEED-EB certification. It was their initiative to work toward that measure of achievement.

The Ten-Point Program

Through the research for this book, I've developed a 10-point program that will help anyone implement a LEED-EB program with greater speed and less cost. These pointers will also help make the certification business case to various stakeholders. This 10-point program can be used as a template for organizing your thinking about LEED-EB, as it is broader than just the details of how to get a project certified (the subjects of Chaps. 7 through 10). The program deals with how to make the case for LEED-EB and how to get buy-in throughout the organization.

Based on the lessons learned in LEED-EB projects by more than 200 successful certifications, this "10-point" program should help you with greening any existing building. Executing each of these elements is crucial for a successful project. Table 11.2 shows the 10 points, and then I describe them in more detail in what follows. This

TABLE 11.2 A 10-POINT PROGRAM FOR GREENING EXISTING BUILDINGS

POINTS	DESCRIPTION
1. Executive leadership	Create a mission/vision statement, clear goals, and strong sustainability policies
2. The LEED-EB task force	Put one person clearly in charge of executing the vision; have a broad makeup
3. The building audit/Gap analysis	Examine options through decision-making focused on the LEED checklist
4. Budget for improvements	Create a realistic budget based on audit results
5. Communications	Communicate rationale for new LEED-EB policies and procedures, internally and externally
6. Knowledge management	Capture "lessons learned" for future projects and project recertification
7. Commitment to continuous improvement	Use LEED-EB as a springboard to continually improve environmental performance
8. Tracking costs and benefits	Use LEED-EB as a catalyst for more closely monitoring the financial implications of upgrades
9. Carbon footprinting	Establish and reduce footprints over time; track carbon emissions reductions
10. Corporate sustainability reporting	Go beyond one building to assessing the sustainability of the entire company's operations

program takes you well beyond LEED-EB and should form the basis for every corporate sustainability initiative.

EXECUTIVE LEADERSHIP—SETTING THE VISION

The question for the CEO and senior management is: Where do you want sustainability and LEED-EB to take you? Is the aim for "zero net impact" overall by a certain date? Is the firm aiming at becoming "carbon neutral?" Is that even possible? If that is the goal, which definition and measurements of carbon-neutral will you use?* The organization's leadership must scout the terrain and chart the course, as well as inspire the team to start down the path and complete the journey.

From his own LEED-EB project experience, Mark Bettin at Chicago's Merchandise Mart agrees with this perspective:

> Certainly having top-down objectives is important especially if you are going to spend money in new areas. So support from the top of the company is critical for LEED projects. Originally, the question about achieving certification was posed to us in the operations group, but it wasn't a question just for operations; the company as a whole was part of the process. It's those all-encompassing objectives that are the key to success. When you have a large company it's always a challenge to get everybody on the same page with the objectives. In that same vein, when you're talking "green" and "sustainability"—those are all-encompassing terms and within that, everybody has limited resources, so in the end, you need to create goals and objectives that work for your organization and have a decent chance of being attainable.

Corporate social responsibility (CSR) or corporate sustainability (CS) commitments can be very important as a guiding principle for LEED-EB programs. Many organizations need to know how the certification program and activities fit into a larger framework. Most of these CSR or CS programs consider all or most of the following areas:

- Climate change, carbon emissions, and direct/indirect energy use
- Water conservation and the associated "water footprint" (especially a manufacturer)
- Waste management, recycling, and product end-of-life disposal
- Suppliers, tenants, customers, and other stakeholders in the organization's activities
- Community impacts, including visitor impacts if open to the public (retailers, universities, schools, recreation centers, etc.)
- Employee wellness and well-being, such as commuting patterns
- Health and safety, including indoor air quality

The involvement of the CEO in the CSR or CS steering committee is critical for reinforcing the corporate vision of sustainable operations as the mission of the company.

*http://online.wsj.com/article/SB123059880241541259.html, accessed July 24, 2009.

The large U.K. retailer Marks & Spencer adopted a "Plan A" in 2005 (because if we all don't achieve sustainability, there is no Plan B as a backup).* Plan A is a visionary approach for a major company: setting 100 targets for sustainability, of which 12 have a direct impact on the built environment. Marks & Spencer wants to become carbon neutral by 2020 and to reduce energy consumption in head offices, stores, and warehouses by 25 percent by 2012. The company prepared a *Sustainable Construction Manual* in 2006, starting with the "Plan A" goals. It then looked at the impacts of carbon, water, waste, materials, biodiversity, travel and access, procurement, and their impact on the built environment. They aligned each of these areas with larger goals for a five-year period and set targets and developed key performance indicators (KPIs) for each. They subdivided each of the targets into incremental annual targets to assist both employees and suppliers in ensuring that they could take bite-sized steps toward reaching the end goals.

At McDonald's LEED Platinum-certified headquarters near Chicago, the company wanted to pursue LEED-EB because of a commitment to sustainability. Helee Hillman of Jones Lang LaSalle, the site manager, emphasized the importance of high-level corporate commitment, because of the many possible stumbling blocks for any major corporate change initiative, such as LEED-EB. As a sign of this commitment, McDonald's has issued a "Global Best of Green 2009" report that highlights 80 case studies of McDonald's operations all over the world engaged in sustainability practices.†

THE TASK FORCE

Once the vision is set, then the real work starts. Typically and effectively, a building or facility task force is formed, with the goal of securing a LEED-EB certification within a 12-month time frame (Fig. 11.1). The task force could include corporate executives or non-facilities employees interested in the larger goal of sustainability for the entire organization. The task force should contain a diversity of skills, viewpoints, roles, and experiences. With strong internal leadership and sometimes an experienced outside LEED consultant, the group needs to develop and deliver recommendations for the required LEED-EB upgrades and renovations recommendations for approval by senior management. Sometimes there will be recommendations and a budget that, say, will deliver a certified or Silver certification, along with some "stretch" activities that might bring the project to the Gold or Platinum level. Be sure to include the associated incremental costs for each certification level. Having experienced people is key. Jones Lang LaSalle's Hillman says:

> Make sure that you have a really good assessment up front by people who didn't just pass their LEED AP test yesterday, but by people who have worked on projects and understand what goes into a project and can quantify and accurately predict which points to attempt. A good project starts off with a good assessment. Also, it's important

---

*Jerry Yudelson, *Sustainable Retail Development*, Chapter 13, 2009.

†www.crmcdonalds.com/publish/csr/home/about/environmental_responsibility/best_of_green.html, accessed May 31, 2009.

Figure 11.1 **To receive the LEED-EB pilot program's first certification, 24 staff in the facilities department at the Getty Museum in Los Angeles spent 2000 hours preparing the required documents.** *Cindy Anderson, © 2003 J. Paul Getty Trust.*

not to over-shoot or over promise to the client, and manage expectations. For example, if the client was promised Platinum by an over-confident assessor who might not know the extent of what goes into each credit for achievement, or you submit a project for Platinum and then you find that the USGBC denies a few credits and you end up Gold, there might be a feeling of disappointment or failure on this project which attains Gold certification, when actually Gold certification is completely respectable and something to be proud of.

Anyone experienced in corporate life knows the importance of managing expectations among senior management.

Staffing the Green Initiative Who's going to do the work? In a typical corporate, industrial, or institutional organization, everyone is working hard and long hours already. Often, someone needs to be hired, or someone from the in-house task force needs to be put into a different position, to act as the LEED project director or corporate sustainability director. This position must have enough clout to get things done, but should not be a separate staff position, removed from the daily work of the company. Many companies continue the consulting relationship beyond the expiration of the initial task force mandate, so that initiatives will be properly advised and the balance of the company's work force can be engaged over a multiyear period.

THE BUILDING AUDIT AND GAP ANALYSIS: EXAMINING GREEN OPTIONS

Now the tough work begins; you carry out the building audit, examine the company's operations, perform a retro-Cx analysis (Fig. 11.2), and look at where you stand versus the LEED credits then, a so-called gap analysis. Then the question arises: to which LEED measures should the building or facility commit? Within these spheres of continuing business activity, there are multiple options to be considered. Should the company focus on basic or higher levels of certification? What about significantly upgrading the energy performance of existing properties, tracking the carbon footprint, increasing purchases of environmentally preferable products, or some combination of all of these? Tenants, suppliers, and customers must also be taken into account; what they are willing to pay for and how they can support the project.

Figure 11.2 **Marriott Corporation has committed its 865,000-square-foot corporate headquarters to LEED-EB Gold certification. Built in 1979, the building expects to reduce water consumption by 20 percent as part of the certification process. Through retrocommissioning of its hotels, Marriott saves $4.5 million annually on energy costs.*** *Courtesy of Marriott International, Inc.*

*Green Building White Paper, *Building Design & Construction,* November 2008, p. 45, www.bdcnetwork.com/article/CA6566572.html, accessed June 2, 2009.

BUDGETING FOR IMPROVEMENTS: ADOPTING SUSTAINABILITY INITIATIVES

At some point, choices must be made and the team must move forward. Most options involve budget considerations, although many consultants told us that a well-designed, well-managed building shouldn't experience a lot of additional capital costs to achieve LEED certification (Fig. 11.3). But seen as a major strategic initiative, sustainability isn't going to be free, even if the major issues are changing some policies and occupant behavior (with respect to recycling and purchasing). For some, especially

Figure 11.3 Park Tower at South Coast Plaza in Costa Mesa, California, received a LEED-EBOM Gold certification in 2009 with a strong emphasis on water conservation, energy efficiency, waste recycling, and green cleaning.
Courtesy of The Offices of South Coast Plaza.

in older buildings, it may be a costly initiative at the beginning to make the necessary changes. Ideally, the suite of initiatives is incorporated into a multiyear plan, with funds set aside for the first three years of activity. This is how GE budgeted the water savings improvements at the Waukesha facility in the Treasure Hunt example earlier in the chapter.

COMMUNICATIONS

Effective communication is a requirement of every serious sustainability effort, but the "story" cannot outrun the achievements, so there has to be a close linkage between the communications function, typically housed in a corporate marketing department, and the operations side of the business. Organizational leadership will probably want to track and report the carbon footprint, issue an annual or biennial sustainability report, take part in conferences and industry forums, have a sustainability or green Web site, and engage in other activities that tell both the internal and external stakeholders what the company is doing and where it expects to go in the future.

In addition to "outbound" marketing communications, there is also a great need for "inbound" tenant or employee communications, to get the most effective cooperation and support for the LEED-EB initiative. Dana Schneider of Jones Lang LaSalle says that tenant engagement is one of the keys to success for the retrofit and ongoing green operations at the Empire State Building:*

> The tenants are really enthusiastic because they're occupying a building that improves the infrastructure that serves them. The building has exceptional outreach to tenants, which is really unparalleled right now. I haven't heard of other buildings doing this where you actually develop plans, sample LEED-CI checklists and design guidelines for tenants to optimize performance in their own space. Part of that includes a tenant energy management program so that tenants will be able to see through a Web-based tool how much energy they're using, how they can improve it and get facts about carbon footprint reduction and energy usage. It's a tremendous plan to engage tenants in what the base building is doing and tying it back to what the tenants are doing and what they can do in their own spaces.

KNOWLEDGE MANAGEMENT/INTERNAL EDUCATION AND TRAINING

The organization needs to establish clear protocols for capturing lessons learned from each LEED project, so that every subsequent project becomes easier, cheaper, and faster. There must also be a commitment to education and training. For many staff, this means literally "going back to school," for example, studying to become a LEED Accredited Professional. For others, it's training in how to apply LEED-EB or LEED-CI to current remodels and renovation projects.

---

*Interview with Dana Schneider, April 2009.

CONTINUOUS IMPROVEMENT

The number of developers and retailers who are using continuous improvement tools is impressive. Alongside LEED-EB, many Europeans may want to use the ISO 14001 environmental management standard.* Tools like this are used to set goals, re-engineer processes, and track progress toward explicit goals. Earlier in the chapter we saw how Gretchen Hancock of GE adapted Japanese lean manufacturing methods to create treasure hunts through continuous improvement methods. Since most people engaged in building and facility management haven't had exposure to lean manufacturing, LEED presents an ideal opportunity for cross-fertilization between building management methods and lean manufacturing practices.

Greening all of your buildings At some point, the outside world is going to ask how a firm is expressing its sustainability values in its buildings. Buildings are often the prime physical manifestation of an organization's environmental performance. People may ask: Are they certified green by some independent third-party? Are they significantly lower in energy use, water use, and waste disposal? Who vouches for their performance? Are the changes real, measurable, and permanent? A firm should be prepared to spend "real money" in the area of sustainability, until all operations are fully overhauled, so that green becomes the norm. For starters, green specifications for refurbishments and tenant improvements should be written and adopted by the firm for use in all of its future projects.

Greening all of your operations There is much to do on the property management side. In the arena of commercial offices, CB Richard Ellis, the largest such property manager, committed more than 200 of its properties to certification through the LEED for Existing Buildings program.

Large owners or managers of retail and commercial office spaces, for example, may want to negotiate national waste recycling contracts, engage with energy service companies (ESCOs) in shared-savings contracts, execute third-party solar energy investments on larger properties, rewrite janitorial and landscape maintenance contracts, work with retailers and utilities to recycle mercury from fluorescent lamps, and engage in a host of other activities that go well beyond compliance with local and national energy and environmental regulations. In this way, implementing LEED-EB measures at one facility can create a springboard for a host of corporate sustainability initiatives across a company's entire building portfolio.

Lean and green Jones Lang LaSalle (JLL) published a sustainability perspective that can be used to guide continuous improvement efforts. JLL calls the approach "lean and green" and identifies three basic principles, as follows:†

---

*www.iso.org/iso/iso_catalogue/management_standards/iso_9000_iso_14000/iso_14000_essentials.htm, accessed July 16, 2009.

†www.joneslanglasalle.com/Lists/ExpertiseInAction/Attachments/110/sustainability-lean-and-mean-means-green.pdf, accessed May 31, 2009.

- *Do what you can for free.* Make sure lights go out at night when everyone is gone; adjust thermostats up in the summer and down in the winter (similar to GSA's strategy); turn off equipment when not in use; adopt daytime cleaning.
- *Consider low-cost, quick-payoff improvements.* Save 5 to 10 percent of energy costs immediately with motion sensors to turn off lights in unoccupied rooms; programmable thermostats to adjust temperatures after hours; task lighting.
- *Budget for longer-term, big-payoff strategies.* Reinvest savings in energy retrofits, creating a "virtuous cycle." Adopt retro-Cx for all buildings you own or manage. Shrink overall office area with alternative workplace strategies, such as shared workspaces, more room for collaborative projects and fewer individual cubicles.

TRACKING COSTS AND BENEFITS

Since the case for LEED is always going to depend on how well the proponents present the business case, for future projects, it's important to track the costs and benefits of LEED project measures. Many of the benefits such as health and productivity can be hard for building management to track, while presenting hard data on utility savings in terms of return on investment is more straightforward.

Case study: Adobe Systems Incorporated, San Jose, California Adobe Systems, a leading software company, certified its three headquarters buildings at LEED-EB Platinum in 2006. Serving about 2300 employees, these structures represent nearly one million square feet of total building area and range in age from 5 to 13 years old. Cushman & Wakefield's general manager and the chief instigator of the LEED-EB-certification effort, George Denise, presented the results of the investment in numerous forums, helping to awaken other companies and project teams to the "low-hanging fruit" available to be plucked by LEED-EB efforts.

According to Denise:*

> Through December of 2006, in preparation for certification, Adobe completed 64 projects, spent approximately $1.4 million on energy conservation and related projects, received $389,000 in rebates from local and state agencies, and reduced annual operating costs by $1.2 million. This is a 9-month payback with a return on investment of 121 percent.

About 69 percent of the total investment costs came from 51 projects involving load management, lighting changes, and equipment retrofits. Adobe achieved some astounding results from this series of LEED-EB projects:

- Electricity use decreased 35 percent per occupant, and natural gas use 41 percent.
- Municipal water use decreased 22 percent, with landscape irrigation water use down 76 percent.
- Carbon dioxide emissions were reduced by 16 percent.

*George Denise, "Adobe Systems Incorporated: A Case Study in Operating Sustainably," 2008, pp. 73-81 in *Lessons Learned: The Costs and Benefits of High-Performance Buildings*, Volume 5.

- Up to 90 percent of solid waste was diverted through composting and recycling.
- About 20 percent of employees use public transit, versus a 4 percent local average.

The company's ENERGY STAR scores were 84, 87, and 93, remarkable considering that the facility ran three data centers and 28 software labs with very high energy intensities (measured in watts per square foot) compared with conventional office buildings.

CARBON FOOTPRINTS

Awareness of carbon footprinting is growing, and many companies are now tracking their carbon footprints as a significant way to measure the results of sustainability programs. *Footprinting* in general is a method of assessing ecological impact by determining the amount of acreage it takes to support human activities.

Carbon emissions are only one component of greenhouse gases, but represent those most frequently tracked by the commercial, industrial, and institutional building sector. The Greenhouse Gas Protocol (GHG Protocol) is a sustainability accounting tool covering the six greenhouse gases covered by the 1997 Kyoto Protocol—carbon dioxide (CO_2), methane (CH_4), nitrous oxide (N_2O), hydrofluorocarbons (HFCs), perfluorocarbons (PFCs), and sulfur hexafluoride (SF_6).* However, at this time, most of the focus is on carbon dioxide emissions from electrical power use, natural gas combustion, and transportation operations.

The GHG Protocol, a multistakeholder partnership, provides standard accounting procedures, data, and calculation tools. It is the most widely used international accounting tool for quantifying greenhouse gas emissions and provides the basic standard for reporting emissions data.† The GHG Protocol contains tools customized for specific industries developed with stakeholders, practitioners, and policy makers from affected industry groups.‡ It also helps companies prepare for other GHG reporting initiatives.

Based on the GHG Protocol, the Climate Registry is the North American standard for reporting GHG inventories§ and prepares companies for new mandatory greenhouse gas reporting standards like California's AB 32 legislation.¶ It is a collaborative effort between more than 40 states, provinces, and tribes in the United States, Canada, and Mexico to develop and manage a common and unified greenhouse gas emissions reporting system.**

---

*http://www.ghgprotocol.org, accessed March 5, 2009.

†Ibid.

‡Ibid.

§Note that there are many new GHG accounting and reporting tools entering the market, including:

- OpenEco (www.openeco.org), a web application and online community for reporting, analyzing, and comparing GHG emissions. OpenEco allows organizations to compare their emissions against other organizations. (OpenEco was developed by Sun Microsystems, in partnership with Natural Logic.)
- California Air Resource Board's GHG Reporting tool (http://www.arb.ca.gov/cc/reporting/ghg-rep/ghg-tool.htm), which will support mandatory reporting in California.
- The California Climate Registry—The Climate Action Registry Reporting Online Tool (CARROT) (http://www.climateregistry.org/tools/carrot.html), California Registry's greenhouse gas emission calculation and reporting software. Participating companies can benchmark their emissions data and public users can view high-level aggregated emissions data.

¶http://www.ghgprotocol.org, op.cit.

**www.theclimateregistry.org, accessed March 5, 2009.

The GHG Protocol has the following objectives:*

- Helps companies prepare a GHG inventory representing a true and fair accounting of emissions through the use of standardized approaches and principles
- Simplifies and reduces the costs of compiling a GHG inventory
- Provides business with information useful for building an effective strategy to manage and reduce GHG emissions
- Provides information facilitating participation in voluntary and mandatory GHG programs
- Increases consistency and transparency in GHG accounting and reporting

GHG calculators The GHG Protocol's calculation tools are presented in Microsoft Excel spreadsheets with the following step-by-step guidance documents:†

- Overview of the protocol (with information for specific business sectors)
- Analytical approaches for determining CO_2 and other GHG emissions
- Guidance on collecting activity data and selecting appropriate emission factors
- Likely sources of GHG emissions and the scopes they fall under (sector specific)
- Quality control practices and program specific information

The GHG Protocol provides separate calculation tools for each industry sector and different tools for input of different data types. Emissions accounting is typically broken down into categories such as direct emissions (fuel used on-site and in company-owned mobile emissions sources), indirect emissions (purchased electricity), and indirect sources (employee commute and business travel). Greenhouse gases are usually reported on a yearly basis, so typical data inputs include utility expenditures and fleet mileage over a year.‡ The GHG Protocol's tools use emission factors (amounts of greenhouse gases emitted by a set amount of business activity) based on data from the Intergovernmental Panel on Climate Change (IPCC).

Refurbishing existing buildings in the United Kingdom A 2009 study by consulting firm Cyril Sweett outlined the potential for reducing GHG emissions through building renovations in the U.K.§ For its analysis, the project identified seven types of offices, along with one generic retail and one industrial/warehouse building. The key findings were:¶

> Modernizing older offices to current market standards reduces baseline CO_2 emissions by about 25 percent. Refurbishing all older offices up to modern standards can be

*http://www.ghgprotocol.org, op.cit.

†ibid.

‡ibid.

§Cyril Sweett, "Costing Energy Efficiency Improvements in Existing Commercial Buildings, Investment Property Forum, Summary Report, January 2009. www.cyrilsweett.com/pdfs/IPF_low_energy_improvements_summary_report.pdf, accessed May 31, 2009.

¶Ibid., p. 13.

achieved with no additional expenditure above the cost of a standard refurbishment. Additional expenditure of $7.60 per square foot (£50/square meter) would reduce total baseline emissions by roughly 50 percent for older offices.

Additional expenditure of $7.60 per square foot when undertaking a refurbishment of 1990s offices would achieve a baseline reduction of 42 percent to 51 percent.

The study also looked at supermarkets and industrial-warehouses and concluded that the energy efficiency of both building types can be improved significantly from just upgrading the buildings to market standards, particularly by spending an additional $1.52 per square foot (£10/square meter) more than the cost of a typical refurbishment. Supermarkets offer a greater opportunity for carbon emission reductions than industrial facilities because of their higher energy use intensity.*

In 2008, the Carbon Trust also published similar guidance for existing "non-domestic" buildings in the United Kingdom. The report concluded, "A key common factor in successful low carbon refurbishments is a corporate commitment to cutting carbon emissions, coupled with effective project management to ensure that this is translated into targets and approaches which ensure the low carbon objectives are met."† Some of the key recommendations in the guide from The Carbon Trust include:‡

- Secure commitment from the senior team by agreeing to low carbon objectives as part of the project vision statement
- Establish the current carbon footprint of the building and set carbon reduction targets for the refurbishment
- Consult building occupants and key stakeholders at the beginning of the process and ensure project buy-in from the design team and site workers
- Appoint a carbon champion at an early stage of the project to maintain a focus on energy use implications of design decisions
- Integrate low carbon design into the general building design and don't treat it as an add-on
- Use a whole-life cost analysis to evaluate low carbon systems and components
- Ensure high quality commissioning for energy efficiency, allocating a specific budget for the purpose

One conclusion to draw: Standards such as LEED-EB (and related green building rating systems in other countries) can be a powerful force in reducing GHG emissions throughout the commercial (non-domestic) building sector. Similar studies in North America would undoubtedly yield similar results and approaches. Investing in energy efficiency and green operational practices in commercial buildings is crucial for controlling human-induced climate change through reducing carbon emissions.

---

*Ibid., p. 14.

†The Carbon Trust, "Low Carbon Refurbishment of Buildings, A guide to achieving carbon savings from refurbishment of non-domestic buildings," Management Guide CTV038, www.carbontrust.co.uk/publications/publicationdetail.htm?productid=CTC751, accessed July 18, 2009.

‡http://www.carbontrust.co.uk/News/presscentre/2008/low-carbon-refurbishment-guide.htm, accessed July 18, 2009.

CORPORATE SUSTAINABILITY REPORTING

Many companies go beyond LEED certification and engage in full-scale sustainability programs, committing to report results according to accepted disclosure formats, such as the Global Reporting Initiative (GRI).* According to the GRI, "Sustainability reporting is the practice of measuring, disclosing, and being accountable to internal and external stakeholders for organizational performance towards the goal of sustainable development." GRI's mission is to provide a credible framework for sustainability reporting that can be used by organizations of any size, industry sector, or geographic location. It represents a globally shared framework of concepts, consistent language, and metrics.

GRI's sustainability reporting guidelines consist of principles that define report content to ensure the quality and consistency of reported information. It also includes standard disclosures, including key performance indicators. Key reporting principles include what GRI describes as "materiality," so that sustainability reporting is not limited only to those sustainability topics that have a significant *financial* impact on the organization. Material issues for a sustainability report might also include *economic, environmental, and social* impacts that cross an impact threshold. Information should also be balanced, comparable with prior periods, accurate, timely, reliable, and clear.

Summary

This chapter presented an original "10-point" program for approaching the greening of existing buildings, focused on the activities of decision makers among property owners, building and facility managers. These 10 points include:

- Executive leadership in creating a mission, clear goals, and sustainability policies
- Organizing the task force
- Examining options through building audits and focused decision-making
- Budgeting for improvements and upgrades
- Internal and external communications, both during the project and afterward
- Knowledge management, how to keep the greening of the building going forward
- Instituting lean thinking and continuous improvement, incorporating innovations
- Tracking green building costs and benefits, especially in energy, water, and waste
- Carbon/water footprint calculations, reductions in emissions and tracking
- Sustainability reporting for the organization, going beyond one building at a time

With sufficient attention to these 10 points, an organization will be well on the way toward sustainable operations, with gains in operating cost savings and employee satisfaction as important early results.

---

*Sustainability Reporting Guidelines, Version 3.0, www.globalreporting.org/ReportingFramework/G3Guidelines, accessed July 16, 2009.

12

GREENING THE FUTURE

I've spent a lot of time in this book showing you why greening existing buildings is relatively easy using the LEED system and almost always cost-effective. Figure 12.1 reviews the many benefits of green buildings; they are there in almost every project, they are real and are beginning to be noticed, as shown by the example of the UBS Tower in Fig. 12.2. There is no doubt that the pace of building energy retrofits and green upgrades is going to accelerate dramatically in the next five years. Certain macroeconomic and regulatory developments such as higher energy prices, building energy labeling, a national energy code, and state-level laws to limit carbon emissions might well accelerate the trend, but it is already well established.

Emerging Trends

What else might be on the horizon? It's entirely possible that there are "weak signals" already in the environment that can forecast momentous trends. One of the classic examples would be the way Starbucks picked up on the trend in the early to mid-1980s toward people wanting to drink good coffee (remember the awful taste of MJB, Folgers, and Maxwell House?) as well as the trend toward one-person businesses that needed a place for millions of independent practitioners to meet. So Starbucks set out to become the nation's "third meeting place" (after home and workplace). No one could have predicted at the time how ubiquitous the iconic Starbucks coffee shop would become and how integrated it would be with daily life for many people.

If you want to stay on top of this rapidly growing industry, you'll have to read everything possibly relevant (and I would add, everything else that relates to design, development, and global warming, for starters); stay on top of trends in every sector of the business; trust your intuition; be open to disruptive ideas and innovations (imagine where the iPhone and 24/7 "always on" access to everyone, everywhere will eventually take us); be aware of your own biases (most people think, "the future will be like the past, only larger or smaller"); and have an ongoing process to identify and

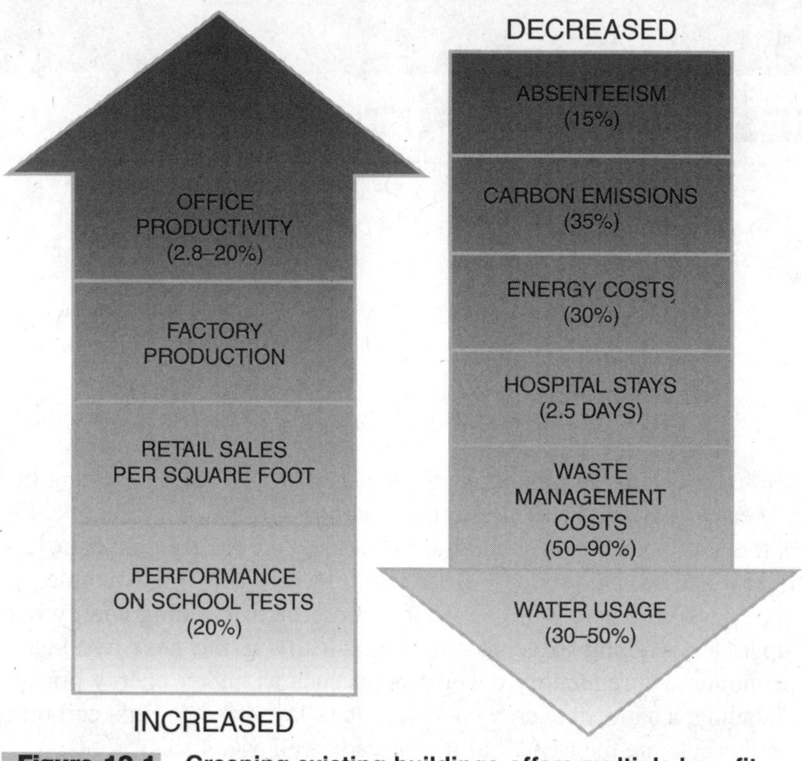

Figure 12.1 Greening existing buildings offers multiple benefits, when other features such as daylighting and indoor air quality upgrades are combined with retrofits to reduce energy and water consumption. *Source: USGBC data, www.usgbc.org.*

monitor weak signals, with the ability to prioritize actions for your company or organization based on what you find out.

What are the weak signals that could spur momentous changes in the green building world? Here are a few that have popped up since 2006:

- The trend in venture capital to focus on renewable energy, carbon mitigation, and energy-efficiency technologies. This could mean that concept-busting innovations might be coming a lot sooner than you think. Consider solar-powered roofing systems that deliver electricity at less than 10 cents per kilowatt-hour and what that could do for innovations in building design and for creating zero-net-energy homes and buildings.
- The availability of software, sensors, and intelligent networks that make buildings so smart they can identify their problems and come up with fixes, so that we can achieve the 90 percent reductions in energy use that are part of every scenario for controlling global warming. Already, we're seeing progressive electric utilities in California aggressively investing in "smart metering" and partnering with leading Internet firms such as Google to monitor and control energy demand in commercial

buildings.* Energy and technology expert Jack McGowan talks about the coming convergence of information technology and building operations:†

Central to this idea is making building systems smarter. It is one thing to tout the use of systems to make the building intelligent, but the driving factor for the integrator and the owner is the human cost of making sure the technology delivers its promise. Using an energy management Web portal and web services like GEMS‡ is just taking another page from the information technology book on how to use systems to their full potential. Companies like Oracle and SAP have built very successful models around this idea of using computing power to point out problems and highlight where performance needs to be improved. The building industry has spent the better part of the last two decades trying to wring engineering and implementation labor and cost out of system installations. Now it is time to wring labor out of the day-to-day tasks associated with assessing the effectiveness of building operations.

- In May 2009, for example, Google announced that it had entered into partnerships with eight energy companies to use its PowerMeter energy management software. Six of the new partners, like San Diego Gas & Electric and Wisconsin Public Service, are in the United States, while two others are in Canada and India. Google's PowerMeter product lets users log into their utility's Web site and use PowerMeter to monitor and manage their energy use, potentially saving money for the customers and reducing demand on the supplier. Will there be any excuse remaining for running buildings inefficiently when you can easily monitor energy usage on a real-time basis?
- In April 2009, Portland General Electric began rolling out more than 800,000 "smart meters" across its 4000-square-mile service area in Oregon after successfully completing its smart metering systems testing program.§ The core idea here is to tie in rooftop solar power generators, office equipment such as printers and copiers, HVAC systems into a seamless grid, so that demand can be managed.
- The continuing trend to cheapen and miniaturize communications technologies, along with an increasingly mobile workforce, means that we can make buildings half the size we think we need and still accommodate workforce and productivity needs. This trend should also allow single-tenant buildings to increase their income by subleasing existing space and/or redesigning workspaces for more collaboration and less individual "cube farm" environments.

---

*http://www.smartgridnews.com/artman/publish/news/Google_Announces_Utility_Partners_for_PowerMeter-587.html, accessed May 31, 2009.

†www.automatedbuildings.com/news/may09/articles/mcgowan/090428101001mcgowan.htm, accessed June 2, 2009.

‡There is a need for a real time management and energy policy enforcement tool for the building owner. The first tool of this type is called global energy management for sustainability (GEMS). The GEMS Web portal allows owners to manage campus sustainability in real time. Converting terabytes of energy and building information is a major challenge and this tool is designed to change data into knowledge. For example, at the University of New Mexico the GEMS Web portal publishes energy data and carbon footprint in real-time. www.automatedbuildings.com, op.cit., accessed June 2, 2009.

§http://tdworld.com/customer_service/pge-smart-meter-installation-0409/, accessed May 31, 2009.

- LED lighting is moving rapidly to replace even compact fluorescents, with half the power consumption. This will allow the further downsizing of mechanical HVAC systems, opening the way for new HVAC technologies to serve buildings' needs, or else accommodating more people in the same space without having to add cooling.
- Nanotechnologies (at the scale of one-millionth of a millimeter) will revolutionize every aspect of building technology, from self-cleaning windows, to self-cleaning façades, windows with 0.1 U-values (R-10) (instead of 0.5, or R-2 now), wall paints that reflect infrared radiation, and so on.
- Building information modeling will enable integrated design to happen in real time, so that you could know the energy use of every possible building design change in minutes, instead of weeks. The modeling will also be used more in large-scale retrofits, as we showed with the Empire State Building, to investigate the most cost-effective groups of technologies for each situation.
- California's historic drought in 2009, for which most cities are adopting strict water conservation regulations, is a preview of what will be the next big challenge for buildings: reducing water demand. The link between energy use and water use is becoming better known and will lead to a stronger linkage between these two resource flows. California's governor imposed a 20 percent cut on urban water agencies in February 2009.* In the future, no building will be exempt from cutting water use, not only in California but also across the country.
- There is no doubt that during the 2010 to 2012 period many cities will begin mandating building energy retrofits for any existing buildings that require permits for renovations, remodels, etc. In April 2009, Los Angeles became the first city to require its own buildings to be renovated and upgraded to LEED standards. Already, more than 20 states do the same for their new buildings, typically to LEED Silver standards and many large cities require private sector commitments to LEED Silver for new buildings. When more than 900 U.S. mayors have signed a climate change commitment to reduce greenhouse gas (GHG) emissions below 1990 levels, do you think that mandates on the private sector can be far behind? Why not get ahead of the curve and start thinking now about how to meet such mandates in a cost-effective manner? Wouldn't that put you ahead of the competition?
- Climate change and dependence on imported energy sources is being seen increasingly as a national security threat. A 2009 report, "Powering America's Defense: Energy and the Risks to National Security," recommends that energy security and climate change goals should be clearly integrated into national security and military planning processes and that the Department of Defense (DoD) should understand its use of energy at all levels of operations, its *carbon bootprint*. The report also recommends that DoD should transform its use of energy at installations through aggressive pursuit of energy efficiency, smart grid technologies, and electrification of its vehicle fleet. These recommendations are as likely to influence the civilian sector as the military sector.†

---

*http://www.huffingtonpost.com/2009/02/27/schwarzenegger-declares-c_n_170693.html, accessed June 2, 2009.
†www.cna.org/nationalsecurity/energy/report/, accessed June 2, 2009.

Figure 12.2 **One North Wacker, commonly referred to as UBS Tower, is a 51-story, modern 1.4 million-square-foot office building in Chicago's West Loop. This top-of-the-line Class A office tower was completed in 2001 and earned a LEED-EB Silver certification in 2008.* Currently owned by a Hines investment fund,† the office tower's modern design, extensive use of glazing, and advanced HVAC and building control systems made LEED-EB certification easier.** *© Jon Miller, Hedrich Blessing; courtesy of Goettsch Partners.*

Within a few years, social trends, technology trends such as iPhones, Blackberries, etc., and the widespread use of Internet 2.0 tools, along with new social media, quite likely will change the future of green building operations in ways we can scarcely predict now. For example, Twitter messages ("tweets") could be sent out from building operators to remind people at 4:55 p.m. to turn off their computers and monitors before going home. Imagine every building having a Facebook page and every building occupant becoming a "friend." Social networks and smart buildings could easily facilitate occupant comfort surveys, instant communications about energy use, encourage recycling, assist people in finding carpools and commuting options, etc.

---

*UBS Tower at One North Wacker Earns Green Certification, Business Wire, Tuesday, February 26 2008, from: http://www.allbusiness.com/banking-finance/financial-markets-investing-investment/7061527-1.html, accessed May 31, 2009.

†http://www.hines.com/property/detail.aspx?id=2168, accessed May 31, 2009.

Planet Wedges

This may seem like an extremely conservative viewpoint, but if we look back 50 years, we will find that we are still, in the year 2010, producing electricity and fueling our factories, cars, offices, and homes with largely the same energy sources as in 1960: oil, gas, and coal. After promoting nuclear power in the 1960s as the answer for diminishing supplies of oil and gas, the United States turned away from it (no new nuclear plants have been constructed since 1974). While some are revisiting this option as a possible solution to carbon emissions, it won't be until after 2015 and probably after 2020 that new nuclear plants begin to show any significant contribution to GHG reduction. In the short run, energy efficiency investments in buildings and homes are the best bet we have for curbing carbon emissions.

In 2004, Princeton University scientists Stephen Pacala and Robert Socolow introduced a new way of thinking about carbon mitigation, investigating how we might eliminate growth in carbon emissions over the next 50 years, using only existing technologies. To avoid a doubling of carbon dioxide concentrations in the earth's atmosphere, with attendant dramatic climate change, they argue that global society needs to act quickly "to deploy low-carbon energy technologies and enhance natural carbon sinks." Their "stabilization wedges" concept illustrates the scale of emissions cuts needed in the future, and provides a common unit for comparing the carbon-mitigating capacities of various energy-saving, renewable energy, and carbon capture and storage technologies (see Fig. 12.3).*

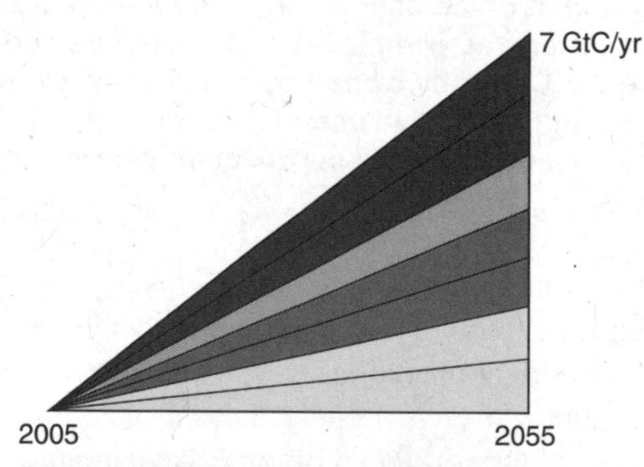

Figure 12.3 Each stabilization wedge represents a reduction of 1 gigaton (1000 million metric tons) of CO_2 emissions per year in the year 2055, if the stabilization measures were fully implemented. Collectively, they would serve to reduce annual global GHG emissions back to 2005 levels.

*http://www.princeton.edu/~cmi/resources/stabwedge.htm, accessed May 31, 2009.

Pacala and Socolow contrasted two futures, the first based on a "do-nothing" premise, that is, continuing with carbon emissions at much the same rate as the past 30 years. The projected result is 14 billion tons per year of carbon emissions by 2056, which would triple "preindustrial" levels of CO_2 in the atmosphere, almost certainly with significant adverse changes in global climate.

In the second, more positive projection, emissions were frozen for 50 years at today's level of 7 billion tons of annual carbon inputs, then reduced by half for the next 50 years, holding total CO_2 concentrations at only double the preindustrial level, less than 50 percent higher than today's atmospheric concentrations.

Socolow and Pacala identified two long-term trends that may help accomplish part of this goal: a transition from primarily manufacturing to primarily service economies all over the world, and the "substitution of cleverness for energy," for example, the development of more energy-efficient appliances and aircraft engines.

To look at solutions, the two professors developed the notion of "stabilization wedges" each of which would avoid one billion tons of carbon emissions per year 50 years from now, starting at zero today. This would move the annual global carbon emissions from 14 billion tons back to 7 billion, as shown by Fig. 12.3.

How does all this relate to greening existing buildings? Pacala and Socolow presented 15 potential ways to get seven stabilization wedges, which would each yield 25 billion tons of total emission reductions over the next 50 years. This is "sustainability thinking" on a planetary scale. One of these wedges requires cutting electricity use in homes, offices, and stores by 25 percent. They said, "The largest carbon savings are in space heating and cooling, water heating, lighting, and electric appliances."* Now double that amount of savings. Think of it: if we could collectively cut 50 percent of energy use in homes, offices, and stores over the next 50 years, that act alone would itself contribute nearly 30 percent (two of the seven wedges) of the emission reductions needed to avert significantly adverse climate change. This is a grand challenge for building owners and managers, facility managers, designers, builders, developers, universities, and government agencies.

The numbers are there for this to happen: The total U.S. building stock equals approximately 300 billion square feet. In the United States every year, approximately 1.75 billion square feet of buildings are torn down. Every year, approximately 5 billion square feet is renovated. Every year, about 5 billion square feet is built new. By the year 2035 (compared with 2005), approximately three-quarters (75 percent) of the built environment will be either new or renovated. According to the nonprofit organization, Architecture 2030, this transformation of the building stock over the next 30 years represents an "historic opportunity for the architecture and building community to avoid dangerous climate change," so there is a chance to dramatically reduce energy use and carbon emissions, one building, one home, and one store at a time.†

*http://www.princeton.edu/~cmi/resources/wedgesumtb.htm, accessed May 31, 2009.

†http://www.architecture2030.org/current_situation/hist_opportunity.html, accessed June 2, 2009.

It is not that hard to imagine cutting energy use in existing buildings by 25 percent over that time frame, and a "piece of cake" to cut electricity use in new buildings by 50 percent over the same time frame. In fact, the AIA (American Institute of Architects)-supported, Architecture 2030 protocol calls for cutting new building energy use by 60 percent by 2010, compared with 2003 averages. Add to that major ramp-ups of wind and solar use in buildings and you've got another wedge, or nearly half the problem solved.

Here's the challenge for building owners, facility managers, engineers, contractors, equipment makers, planners, lenders, investors, and everyone involved in the building industry. Look at your building portfolio: if it doesn't use electricity at least 25 percent below the U.S. average for comparable buildings, go back and recommission it to make it happen. If it's not cutting electricity use by at least 50 percent below the average within three years, consider whether your building portfolio will be competitive five years from now. Can you get upgrades to produce high-performance buildings into next year's budget? If not, figure out a way to do it (Fig. 12.4). There are always low-cost measures that result from any retro-Cx activity, along with a variety of new financing mechanisms profiled earlier in this book, such as energy service companies (ESCOs), the Clinton Climate Initiative programs, and local energy bonds.

The twenty-first century is shaping up to be an "efficiency century." We already know that the cheapest (and cleanest) form of new energy is conservation, or "negawatts,"* and that we need to pursue those opportunities as fast as possible. We should all be encouraging fresh approaches to building and community design and operations that reduce energy use while maintaining high levels of environmental quality and user satisfaction.

The first step in getting better building performance will undoubtedly rest with the more technical side of the business: the mechanical engineers and code officials who are actively coming up with ever more efficient codes and standards. While a detailed exposition of all pending code changes is well beyond the scope of this book, it is worth pointing out the International Green Construction Code (IGCC) initiative, unveiled in June of 2009 by the International Code Council, along with the American Institute of Architects (AIA) and the American Society for Testing and Materials (ASTM International). The purpose of the initiative is to outline a model for improving the design and performance of both new and existing commercial buildings.† The draft standard, expected in 2010, aims at creating a legally enforceable green code that will serve to complement established rating systems such as LEED. As with all codes, however, they serve to create important floors on future building performance, but do not by themselves bring about the rapid reduction in building energy use that will be required to reduce carbon dioxide emissions dramatically.

---

*www.ccnr.org/amory.html, accessed May 31, 2009.

†http://aec.ihs.com/news/2009/icc-green-commercial-building-code-070209.htm, accessed July 18, 2009.

Figure 12.4 VSP® Vision Care received LEED-EB Platinum certification in 2008 for the first of four buildings on its Rancho Cordova, California, corporate campus, reducing waste generation by 70 percent and water consumption by 50 percent.* *Photo courtesy of VSP/Rancho Cordova, CA.*

Net-Zero Energy (Zero-Carbon) Buildings

The United Kingdom is at the forefront of zero-carbon regulation, something I learned in 2008 in preparing a book on European green building trends. For example, by 2016 all new social housing in the U.K. is supposed to be built to zero-carbon standards (called Code Six), with at least 20 percent of the annual energy supply coming from on-site renewable sources. In 2008, the first Code Five apartments were built in Surrey with a cost premium of 24 percent, about $80,000 per unit. The homes include an average of 8.5 photovoltaic modules at a cost of $17,000 per home and a wood-pellet biomass boiler costing about $11,000 per unit.†

The U.K. government includes biomass boilers in the definition of renewable energy sources, but neglects to include in the calculation the transport energy costs of

*https://www.vsp.com/newsroom/html/leed-platinum.jsp, accessed June 2, 2009.

†Building magazine supplement, "The Rules of Engagement," April 2008, www.building.co.uk, accessed May 31, 2009.

the wood pellet supply. I think pellet boilers and wood stoves are climate-neutral for some of the colder and forested regions of the United States, especially those currently relying on fuel oil. They might even be practical in smaller commercial buildings in those same areas, but are probably not a practical solution for most of the country, especially in areas where natural gas is readily available.

Now British attention is turning to developing a similar "zero carbon" requirement for new commercial buildings by 2019. One of the key issues is that nonresidential buildings are far more diverse than homes, ranging from short and tall offices, to shopping centers and malls, restaurants, hotels, churches, recreation centers, data centers, factories, schools, laboratories, hospitals, and standalone stores such as a Wal-Mart or Best Buy. You can imagine the difficulty of writing a zero carbon set of rules for such a diverse range of building types, sizes and end uses, in each of the eight major climate regions of the United States. Given the three-to-five year cycle for planning, budgeting, permitting, building, and occupying major structures in the United Kingdom; this 2019 regulatory date effectively means that all issues have to be resolved by 2015 at the latest, less than six years from now.

Nonetheless, the low carbon future is something that will hit us all quite soon, with President Barack Obama talking about global warming and cap-and-trade legislation likely to be enacted by a heavily Democratic Congress in 2009. What does zero net energy or zero carbon really mean? While these two terms are not quite interchangeable, here are some possible definitions encompassing both, each a bit more confining, as shown in Table 12.1.

What energy demands do zero carbon standards address? Most of the thinking to date has been about heating, hot water, lights, and cooling. But in many homes and

TABLE 12.1 ZERO NET ENERGY/NET ZERO CARBON DEFINITIONS (FROM MORE TO LESS STRINGENT)

1. Produce on-site annually whatever net amount of energy you use, based on *source* energy calculation (about 2.5 times greater than *site* energy), with renewable energy on-site. [One variation for true believers: Never buy any power from off-site (implies seasonal energy storage in northern latitudes).]
2. Produce on-site annually whatever energy you use on-site, based on *site* energy calculation. Make up the difference with renewable energy on-site.
3. Definitions (1) or (2), but with renewable power generated off-site and physically connected to the site through the electricity grid.
4. Definitions (1) or (2), but use green power generated off-site and not physically connected to the site, through the purchase of "green energy" from reliable sources.
5. Definition (4) but purchase only carbon offsets (renewable energy certificates), instead of the actual power generated. Offsets can come from anywhere but must be carefully documented.
6. Only count renewable energy on a net energy basis, excluding manufacturing, transportation, and other life cycle costs.

commercial buildings, other uses of electricity particularly for appliances and electronics can account for a quarter to more than half the total energy demand, so one can't realistically exclude them. But in commercial buildings, the so-called "process loads" vary considerably from one type to another.

Should homes and buildings each have zero net carbon requirements or can some places that generate more renewable energy than they consume help offset those that don't? For example, can I pay my neighbor to put more PV on his roof if mine is shaded? (One does have to acknowledge the positive benefits of trees and shade in urban environments even if they on occasion may hinder renewable energy production.)

Does the net carbon requirement apply to "source energy" or only "site energy?" If my use is 15 kilowatthours per year per square foot, do I have to offset the 50 kilowatthours [170,000 British thermal units (Btu)] of fossil energy it took to produce it? If the answer is "source energy" and the efficiency of the overall grid is about 30 percent, then I'll have to offset more than 3 times my actual energy use to have a clean slate. Using source energy as the criterion for "zero net energy" also makes a powerful argument for on-site energy production using solar, cogeneration or microturbines, so that a building could use the thermal energy from fossil sources that would otherwise be wasted at a distant power plant. The Europeans have come down squarely on the side of using the source energy basis for net zero, or as they call it, "primary energy."*

Now consider energy supply: does a zero carbon energy source have to be on-site, nearby (e.g., connected by a wire to the end use), or can it be off-site completely (such as energy from large PV and solar thermal electric arrays)? Can we have a total annual net zero carbon, for example, with fair weather solar, while acknowledging that we'll have a requirement for "dirty carbon" sources in foul weather? Drawing the boundary is difficult, as one can see now in the debate over carbon offsets.† Does the replacement energy source even have to be in North America, if it's cheaper for me to buy renewable energy produced in Brazil or Indonesia? After all, climate change is a global problem that argues for both local and global solutions.

Should we count biomass power because it takes lots of fossil fuel to produce the biomass feedstock (other than from forest culling, in which case there's still the transportation cost)? What about the concern that biomass diverts agricultural resources from feeding people?

Could some of the total cost of achieving zero-carbon commercial buildings, one by one, be more effectively spent in deploying large-scale renewable energy generation or other carbon mitigation strategies, such as sequestering carbon dioxide generated by coal-fired power plants? After all, any economy has only so much money to invest in building energy conservation and renewable energy supply, without pinching other social needs.

The answers to these questions are complex and critical, but they are well beyond the scope of this book. Nevertheless, the movement toward greening existing buildings will eventually have to confront them.

---

*Jerry Yudelson, *Green Building Trends: Europe*, Chapter 3.

†http://carbonfund.blogspot.com/2008/01/ftc-npr-take-up-carbon-offset-debate_04.html, accessed June 3, 2009.

THE COMMERCIAL BUILDING INITIATIVE

The U.S. Department of Energy (DOE) started a Zero Net Energy Commercial Building Initiative (CBI) in 2008, which is engaged in wrestling with these issues, among other activities, though DOE takes the site energy approach to the definition: "In general, a net-zero energy building produces as much energy as it uses over the course of a year. Net-zero energy buildings must be very energy efficient. The remaining low energy needs are typically met with onsite renewable energy."* Eventually, the logic of our climate change problem and the strong European example will force DOE to revise its definition (and rather soon) to focus on source energy, rather than site energy. Otherwise, one can't even come close to formulating an approach to solving the carbon challenge.

COMMERCIAL REAL ESTATE BEGINS TO HIGHLIGHT ENERGY EFFICIENCY

Associated with the CBI is a series of partnerships, such as the Commercial Real Estate Energy Alliance with representation from a number of large companies engaged in commercial building development, ownership, and management.† According to the DOE, the Alliance has several important aims, among them:‡

- Provide ready access to advanced technologies and analytical tools developed by the DOE and its national laboratories
- Create and share successful, evidence-based strategies for integrating advanced, high-performance technologies into commercial buildings (Fig. 12.5)
- Provide a consistent voice to encourage national manufacturers and distributors to develop and supply efficient products and services to the commercial real estate sector
- Validate the commercial real estate sector's energy and carbon reduction efforts to audiences such as prospective tenants and buyers and the financial community

In early 2009, commercial real estate services company Cushman & Wakefield, with more than 15,000 worldwide employees, signed a memorandum of understanding with the U.S. Environmental Protection Agency (EPA) to reduce the environmental impact of the more than 3200 offices and buildings the firm manages in the United States.§

The voluntary agreement made Cushman & Wakefield the first firm of its kind to embark on an EPA partnership to green properties in the commercial real sector. The firm's goals under the arrangement include reducing energy consumption in its managed U.S. properties by 30 percent by 2012, tracking water usage and promoting conservation, as well as focusing on reuse of industrial materials and sustainable landscaping practices. Under the agreement, Cushman & Wakefield also promised to make energy efficiency a top priority in operating its corporate sites, selecting new offices,

---

*www1.eere.energy.gov/buildings/commercial_initiative/zero_energy_definitions.html, accessed May 31, 2009.
†http://www1.eere.energy.gov/buildings/real_estate/, accessed May 31, 2009.
‡http://www1.eere.energy.gov/buildings/real_estate/goals.html, accessed July 16, 2009.
§http://www.greenerbuildings.com/news/2009/01/09/cushman-wakefield-epa, accessed May 31, 2009.

Figure 12.5 Armstrong World Industries' corporate headquarters building received LEED-EB Platinum certification in 2007. Built in 1998, the building gets 75 percent of its power from wind energy, has cut water use 47 percent, and recycles 60 percent of its waste.* *Armstrong World Industries, Inc., Corporate Headquarters, Lancaster, PA.*

and in seeking green building certification for those spaces when feasible. Cushman & Wakefield will report its progress to the EPA every six months.

Among the leading property management companies committed to LEED-EB, CB Richard Ellis (CBRE) announced in June of 2009 that it had registered 225 projects for eventual certification. This is significant, since CBRE only manages the buildings, yet the company thinks it vital that they be green-certified. Owned by more than 55 different investment groups, the buildings are located in 21 states, mostly in the central business districts of large cities. They represent more than 57 million square feet of floor space, with an average size of 253,000 square feet.†

The Living Building Challenge

Beyond net-zero energy, what about net-zero potable water use and net-zero wastewater generation performance metrics? What about going beyond LEED and only building and operating buildings that have no impact (what many call being "positively good"

*www.armstrong.com/commflrpac/asia1/ep/kr/highlight39712.html, accessed May 31, 2009.

†www.costar.com/News/Article.aspx?id=43E6542B71D78DBC12BD62D6F4EBD546&ref=1&src=rss, accessed July 16, 2009.

instead of "less bad" for the planet). The Living Building Challenge complements LEED by offering a set of performance metrics that point toward the design and operation of "regenerative" buildings, which give back more than they take from the environment.* While it is still a standard in development, the Living Building Challenge is likely to change the discussion in the building industry dramatically in a relatively short period.

Administered by the International Living Building Institute, the Living Building Challenge requires only 16 prerequisites, with no discretionary points: you either are a Living Building or you aren't, it's that simple. The prerequisites include zero-net energy, net-zero water, and net-zero wastewater performance on an annual basis, as well as aggressive material requirements addressing both the significant embedded toxicity and carbon footprint of our current supply chain. Clark Brockman is an associate principal and the director of sustainability resources at SERA Architects in Portland, Oregon, heavily engaged with designing living buildings; his perspective is:†

> I think LEED-EB will completely reinvent how we live and work in buildings in the future. There will be a tremendous number of "lessons learned" coming out of the LEED-EB world that will then inform a design team's efforts in attaining true net-zero performance in new buildings. This is particularly true because the early work on Living Buildings clearly showed that all of the passive and active green techniques together will not make a net-zero building if the people inside and those operating the building do not buy into the net-zero goals. Data coming from LEED-EB projects will greatly inform this occupant/operator dynamic in green buildings.
>
> For any existing building and LEED-EB project, the first thing to have is a real quality benchmark of performance, which LEED-EB requires. Most projects will likely benchmark against the CBECS (U.S. DOE's Commercial Buildings Energy Consumption Survey) database, which averages actual building performance across the country, for given building types. Once you've established those benchmarks, you can really start to determine pathways to net-zero for that building in its particular climate.
>
> The difference between the Living Building Challenge and LEED-EB is that LEED does a great job of telling you where you are at with your current building and its operations, and what the likely next steps are for improved performance. The Living Building Challenge is focused more on the end game, the level of performance for which everyone is ultimately striving.
>
> LEED-EB is a great and powerful tool that is going to see dramatic uptake in the coming years, but it is by its nature a tool for market transformation through incremental improvement. In contrast, the Living Building Challenge places a clear and very challenging stake in the ground pointing in the direction the green building movement will need to go for the foreseeable future, constantly evolving to get us closer and closer to truly regenerative buildings.

---

*http://ilbi.org/the-standard/version-1-3, accessed June 2, 2009.

†Interview with Clark Brockman, May 2009.

If our overall goal is to go beyond being "less bad," then we must do positive good. In this respect, the Living Building Challenge offers a beacon of hope and a well-defined way to proceed. Brockman says:

> By putting out the goal of, "the building wants to be net-zero energy and/or net-zero water," you invert the problem-solving associated with the conventional design process. Net-zero performance on an annual basis is as unambiguous as it is dramatic—and it changes everything about how one designs. Instead of specifying the most efficient gas-fired furnace, you realize, "We can't even have a gas-fired furnace." Or instead of retaining an existing ventilation system in a building, the team is likely to conclude that, "We know for a fact that using air to push heating and cooling around the building is the least efficient way to do it—there must be a better way for us to provide that same functionality in this existing building (e.g., with radiant heating and cooling)."
>
> A great example of this exact thought process in an existing building is the newly renovated Shattuck Hall building on the Portland State University campus—now home to the University's architecture program.* It is a beautifully restored historic building with this "state-of-the-shelf" heating and cooling system. The engineers decided to take out all of the air systems and specified a completely new radiant hydronic (water-based) system, using ceiling paddle fans to provide additional air circulation. They also replaced the windows to improve the building envelope performance, creating a wonderful LEED Gold building. In this renovation project, the design team didn't just incrementally improve the systems in the building—they actually removed whole systems and replaced them with more efficient systems. I think the Living Building Challenge will push people to do similar things in renovation projects to come.

Earlier in this book, I described a similar phenomenon at the Empire State Building. You have to upgrade the windows' energy performance dramatically to decrease energy demand. You also need more efficient mechanical systems in any good energy retrofit, but you need to take the measures in the right order.

Planning for Larger Scale Developments

Most of this book has been focused on greening existing buildings, one building at a time. But an emerging planning science, based on applying the wedge concept at a "micro" level, is tackling larger developments, such as college campuses, urban districts, and mixed-use development. Pioneered by Mithun architects and planners in Seattle's South Lake Union District, the approach was recently introduced into the planning for

*Credits for the Shattuck Hall renovation go to Portland-based SRG Partnership for the architectural design and PAE Consulting Engineers for mechanical, electrical and plumbing design.

the future expansion of Seattle University, a Jesuit institution with undergraduate and graduate programs. Brodie Bain is a principal architect and planner in charge of Mithun's approach for Seattle University's Sustainable Master Plan. She says:*

> Our thinking for the Master Plan was to try to be as comprehensive as possible, with a couple of different approaches. One approach is by not just focusing only on individual buildings. Because this is a campus, we're looking at the broader campus environment and its overall resource impacts. The other piece is addressing the idea of the *triple bottom line*, the three "P's." That involved looking at issues related not only to the planet, but also to people and prosperity. For sustainability to be truly effective on campuses, an institution must be able to continue to prosper, to meet its mission and to support the well-being of the community. The focus of sustainability has to go beyond just natural systems, by also looking at the economics as well as the human-side of things.

Mithun identified five elements to consider in assessing campus-wide sustainability. These elements could easily form the basis for the key performance indicators (KPIs) of a sustainability reporting system. In Chap. 11, we recommended strongly that organizations and companies measure and report on environmental performance. Mithun's five KPIs are:

- Energy and GHG emissions
- Water cycle
- Ecological system health
- Human well-being
- Economic viability

In each of these areas, Mithun assesses the current conditions as a baseline for thinking about the future. Bain says:

> We looked at each element to get a sense of their current situation so we could start to think about some specific strategies for each of them in terms of reducing the university's impact. We isolated the elements in order to consider them separately and establish benchmarking and metrics for each one. This is fairly easy to do for energy, water and carbon but much more challenging for things like ecological systems and human well-being. In every case, we took the idea that the first thing you want to do is reduce demand. The second thing you want to do is switch to green strategies. The third and very last thing to do is offset. We looked at strategies from that perspective. The first question we asked was, "What can we do to reduce demand for each of these different elements?"

Building owners and facility managers wanting to green their portfolios could all take this approach: first reduce demand (similar to the physician's dictum, "primum

*Interview with Brodie Bain, May 2009.

non nocere," first do no harm); then switch to green strategies (e.g., using recycled rainwater or gray water); finally, as needed and only in the requisite amounts, buy green power offsets or use municipal potable water to supply the remaining demand.

Seattle University is planning to double its square footage over the course of the plan. In doing this, their level of resource use will double as well if they continue to operate business as usual. We had to look at the impact of the whole range of strategies on each measurable element—whether it's energy or water or carbon—to convey to the university the range of possibilities and to illustrate the types of overall goals that might be achievable. That is the idea behind the wedge diagram (shown in Fig. 12.6).

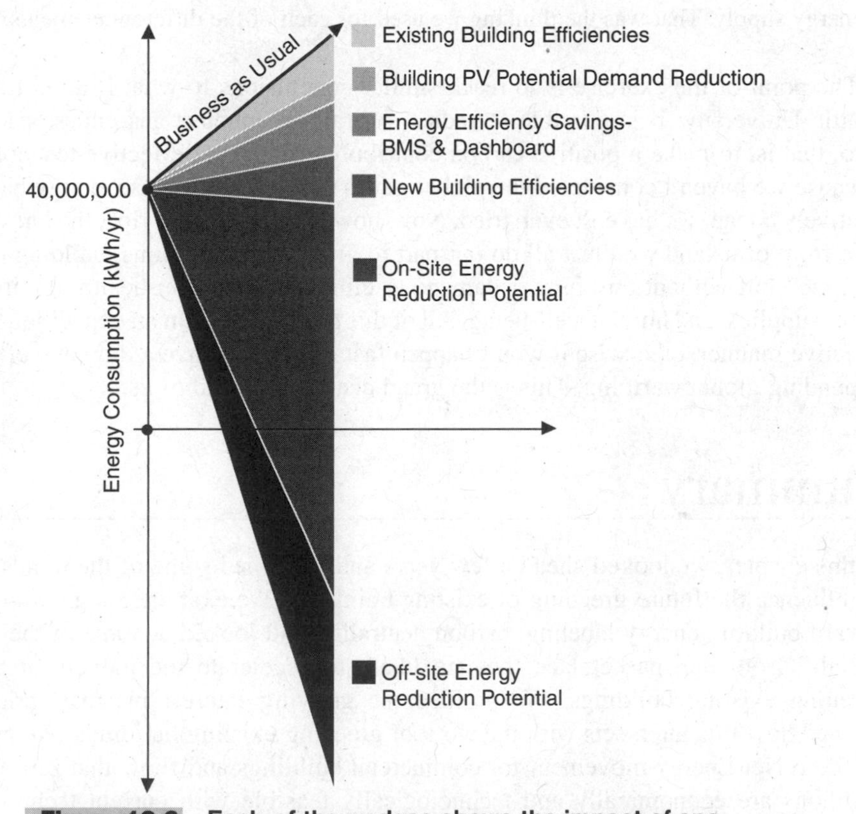

Figure 12.6 Each of the wedges shows the impact of specific strategies on resource reduction at the end of the planning period, ranging from those that reduce demand at the top (these are also generally the most cost-effective) to more intensive green strategies (typically the most financially challenging) at the bottom. The energy wedge diagram shows that theoretically, if the university employed all strategies shown, they could generate twice as much energy as they need. This visual tool helped the university to consider energy reduction goals more confidently. © *Mithun, Inc.*

For reducing carbon emissions, Mithun looked at the same strategies as they did for energy but on a different scale. Each strategy has a different level of impact on carbon than it does on energy.

> The first wedge considers reducing demand for energy by upgrading the existing buildings. The next two wedges also relate to reducing demand by installing PVs on roofs and making people more aware of their impact through education and building monitoring systems. Together, all of those things could reduce the carbon emissions by almost a third. Next is making sure your new buildings are efficient and looking at your campus-wide energy supply system, maybe using combined heat and power instead of a dedicated steam plant. Finally, off-site energy options—purchasing wind power and those kinds of things from other entities can be viable alternatives for energy supply. That was the thinking we used for each of the different elements.

The point of the exercise is to reach similar conclusions to what Mithun found at Seattle University: it is possible to reduce overall greenhouse gas emissions below zero, that is, to make a positive carbon contribution with cost-effective technologies. Because we haven't considered the global issues before and because energy has been relatively cheap, we haven't even tried. Now, however, the time is ripe, the knowledge is in front of us and we must all do our part to grow our institutions, building portfolios, etc., but without any further damage to climate, ecosystem health, diminishing water supplies, and human well-being. All of this must be done in an expeditious, cost-effective manner, otherwise it won't happen fast enough to have a dramatic effect on impending global warming. This is the grand challenge ahead of us.

Summary

In this chapter, we looked ahead a few years and identified some of the trends likely to influence the future greening of existing buildings. We examined a growing trend toward building energy labeling, carbon neutrality, and looked at some of the "weak signals" from the marketplace that are likely to accelerate the movement toward greening existing buildings. We profiled the growing interest in smart grids and showed how this intersects with the work of greening existing buildings. We profiled the Zero Net Energy movement for commercial buildings and show that zero-carbon buildings are economically and technologically feasible with current technologies. Finally, we looked at sustainable retrofits from a broader perspective, focusing on work the Mithun firm has done in sustainable planning for campuses and urban districts. Ultimately, urban planning is a key to long-run sustainability.

13

CASE STUDIES

We're coming to the end of our journey through the fast-growing world of greening existing buildings and you can see that this is a major market trend, one that has emerged largely through the actions of private-sector building owners and managers wanting to reduce energy costs and to certify their own green operating practices. Some of the projects were easy to certify, with almost no additional cost. Other projects, with teams that had little previous experience in green building practices, took considerable time, money, and staff hours. Almost every project that used an outside consultant benefited from the added expertise. With more than 2600 LEED-EB registered projects underway (beyond the 230 or so certified as of May 2009*), considerably more data on the costs and difficulties, as well as the benefits of certification, will be generated in the next few years. In the meantime, we will examine some of the current project registration and certification numbers and profile a few more projects in depth. One thing is clear: LEED-EB has been a "game changer" for building operations, moving them well beyond energy savings to considering the environmental, social, and economic benefits of sustainable sites, water efficiency, materials and resource conservation and indoor environmental quality. The slogan for building owners and operators now must be, "first blue (for energy efficiency), then most certainly green."

LEED-EB Certified Project Analysis

LEED-EB projects come in all shapes and sizes and all levels of certification. Table 13.1 shows the certification levels these projects have attained, beginning with the LEED-EB 1.0 pilot program in 2004, continuing with the LEED-EB version 2.0 program in 2006, and concluding with the LEED-EBOM rating system introduced in 2008.† A detailed listing of all certified projects through the end of May 2009 is contained in Appendix I,

*USGBC data furnished to the author, June 2009.

†The LEED-EBOM 2009 revisions kept almost all credits from the 2008 version intact, but changed the weightings assigned to the individual credits.

TABLE 13.1 LEED-EB RATING SYSTEMS AND CERTIFICATION LEVELS*

	CERTIFICATION LEVEL, %				
LEED EB VERSION	CERTIFIED	SILVER	GOLD	PLATINUM	TOTAL
1.0 Pilots only	24	20	52	4	25
2.0	24	35	27	14	106
O&M	27	33	33	7	30
Grand Total	24	32	32	11	161

*http://www.usgbc.org/LEED/Project/CertifiedProjectList.aspx, accessed May 29, 2009. List is current as of the end of May 2009.
Note: LEED 2.0 certified totals exclude (51) Stop & Shop stores.

"Certified LEED-ED Projects (All Versions)." One thing is noteworthy: except for the Platinum projects, the certification levels are spread out fairly evenly between the first three tiers of certification. (Excluded from the analysis are the 51 grocery stores certified by one owner, Stop & Shop, under the USGBC's "Portfolio" program; these were almost all at the certified level.)

Who is certifying these projects, public or private entities? The answer is shown in Table 13.2. You can see that 83 percent of the project certifications to date have been by private owners, and that about 42 percent of total certifications were at the basic certified level. Public agencies, nonprofits, and educational institutions own about 15 percent of the projects. The average project size is more than 300,000 square feet, which would represent about a 10-story office building in a typical urban location. With a project like this, particularly in a newer building, certification is not difficult

TABLE 13.2 LEED-EB RATING SYSTEMS AND TYPE OF OWNERSHIP

	CERTIFICATION LEVEL					
OWNERSHIP TYPE	CERTIFIED	SILVER	GOLD	PLATINUM	TOTAL	PERCENTAGE
Private	78*	37	46	14	175	83%
Public	7	4	4	3	18	8%
Education	2	6	1	0	9	4%
Nonprofit	2	2	1	1	6	3%
Unknown	1	3	0	0	4	2%
Total	90	52	52	18	212	100%

*Includes 51 Stop & Shop grocery stores.

according to an experienced LEED-EB consultant Craig Sheehy of Envision Realty Services:

> Most of the buildings at this point that we're working on are Class A and Class B commercial buildings that have been pretty well taken care of, or there is already a plan to improve them. That certainly makes the LEED-EB process easier. We've got some buildings that were built in the mid-60s that have been taken care of over time; now we're simply coming in and applying the green practices and principles to them that match up with LEED. They didn't necessarily know that they were aligning with LEED at the time [they made necessary improvements], so it doesn't have to be a brand new building; it can be an older building.

It's becoming quite clear that private building owners see marketing, financial, and corporate sustainability benefits in pursuing LEED-EB projects, even in multitenant occupancies in which operating costs are passed on to tenants. CB Richard Ellis (CBRE) has been a leader in promoting the use of LEED-EB among large property managers, with more than 200 projects going through the program at the time this book was written (see Chap. 12). Gary Thomas is the director of sustainability programs at CBRE. He confirms this impression:*

> The initial buildings going through LEED-EB are generally the best projects in the marketplace. The top-tier sites typically have high ENERGY STAR scores and have implemented some of the [LEED-required] green building practices. They aren't always the newest buildings but they are high performing, high-quality, well-tenanted buildings, so it can make the process a little easier to achieve LEED certification. It's the trophy-type properties that many clients are taking through the program right now, and the next wave of LEED-EB buildings are those that are going to take more time and effort to get through the process.

Other institutions, such as higher education and public agencies, which pay all the operating costs and reap all the benefits of greening existing buildings, have yet to jump onto the LEED-EB bandwagon.

CBRE's Thomas is particularly concerned that many government agencies which lease space in privately owned buildings will start demanding LEED-EB certification as a condition of tenancy, and that will change the market dynamics dramatically. He says:

> Potential future government mandates are also a motivator for building owners to pursue LEED-EB. If the GSA builds out new space, it has to be LEED-certified but as far as leased space in existing buildings, it's a guideline at this point. The perception is that the GSA and all federal government-leased space will, at some point mandate LEED-certification. We have many clients [buildings] with GSA tenants in place and the risk is that if the building owner doesn't take steps to gain LEED certification

*Interview with Gary Thomas, May 2009.

> when it comes time for the lease to roll over, those tenants may be forced to look [elsewhere] for LEED space. In California, for example, where the state government is pushing hard to have state agencies look toward more sustainable office space, our clients with buildings in Sacramento are moving forward with LEED-EB certification for this exact reason.

In terms of geographic location, almost 70 percent of all certified projects are in four states: California, Massachusetts, New York, and Colorado. Since most of the LEED-EB certified Stop & Shop grocery stores are in Massachusetts and New York, it's not surprising that their totals were high. Other states in the top 10, by number of certified projects, were, in order, Oregon, Illinois, Washington, Wisconsin, New Jersey, and Connecticut. Those 10 states contributed more than 70 percent of all certified projects.

LEED-EBOM REGISTERED PROJECT ANALYSIS

Nearly 1400 new projects have registered for LEED-EBOM in the past 12 months, through May 2009, more than doubling the number of registered LEED-EB projects. What all the consultants and project managers are telling us must be true: suddenly the word is out that a well-managed building can get a LEED-EB certification without too much trouble. Tables 13.3 and 13.4 show the registered projects by owner and by the top 10 states.

The percentage of private owners between all LEED-EB certified projects and newly registered projects under LEED-EBOM has remained unchanged, indicating that private building owners, of both single-tenant (such as corporate headquarters or leased to government) and multitenant (large urban offices) buildings, have seen LEED-EBOM as an appropriate and cost-effective way to acquire a "green brand" for an existing property.

What has changed is shown in Table 13.4: the top 10 states, in order of the number of registered projects. California continues to lead the way, but now Texas, Florida, North Carolina, and Maryland are coming up quickly in the ranks. Colorado, Oregon,

TABLE 13.3 LEED-EBOM REGISTERED PROJECTS, JUNE 2008 TO MAY 2009, BY OWNER TYPE*

OWNER TYPE	NO. OF REGISTERED PROJECTS	PERCENTAGE
Private	1159	83
Public	138	10
Education	65	5
Other/Unknown	37	3

*USGBC data furnished to the author, May 2009; author's analysis.

TABLE 13.4 TOP 10 STATES FOR LEED-EBOM REGISTRATIONS, THROUGH MAY 2009

STATE	NUMBER OF LEED-EBOM REGISTERED PROJECTS	PERCENTAGE OF TOTAL PROJECT REGISTRATIONS
1. California	309	22
2. Texas	150	11
3. Florida	87	6
4. New York	79	6
5. Illinois	68	5
6. North Carolina	60	4
7. Maryland	58	4
8. Massachusetts	57	4
9. Georgia	51	4
10. Washington	48	3

Wisconsin, New Jersey, and Connecticut have dropped out of the top 10. As before, the top 10 states have about 70 percent of the total number of registered projects.

Case Studies

LEED-EB certification at any level is a significant achievement and a statement that a building owner is committed to environmental responsibility, to continuous improvement, and that the owner is willing to subject its building operations to outside scrutiny. A commercial building owner leasing space to multiple tenants may also gain a competitive advantage in a time of rapid market convergence to green buildings. Elaine Aye of Green Building Services says:

> Certifications for buildings have become very competitive in the sense of demonstrating to the public that you operate an efficient building. What we're finding with a lot of our Class A office buildings is they're already run very efficiently. They have good engineering. They have good building systems in place, and they've made the necessary upgrades over time. What they need to do is just demonstrate that they're following through. By undergoing certification, it's a way that they can demonstrate that they already have good practices and that they're going to prove it by getting certified. So they're now able to compete on a broader basis.

Sometimes it helps to look at case studies of specific LEED-EB projects, to see some of the common characteristics, as well as to understand the differences in

TABLE 13.5 CASE STUDY CHARACTERISTICS

CASE STUDY	OWNER	LOCATION	CERTIFICATION	YEAR CERTIFIED
1. Sebesta Blomberg	Sebesta Blomberg, Inc.	Roseville, Minnesota	LEED-EB 2.0 Silver	2007
2. The Crestwood Office Building	Melaver, Inc.	Atlanta, Georgia, suburbs	LEED-EB 2.0 Certified	2007
3. Alliance Center	Alliance for a Sustainable Colorado	Denver, Colorado	LEED-EB 2.0 Gold	2007
4. Byron G. Rogers U.S. Courthouse	U.S. General Services Administration	Denver, Colorado	LEED-EB 1.0 (Pilot) Gold	2006
5. San Jose City Hall	City of San Jose	San Jose, California	LEED-EBOM Platinum	2009
6. Bell Trinity Square (483 Bay Street)	Union Real Estate Investment	Toronto, Ontario	LEED-EB 2.0 Gold	2009
7. One Potomac Yard	JPMorgan (leased to U.S. EPA)	Arlington, Virginia	LEED-EB 2.0 Gold	2008
8. 17th Street Plaza	Operated by Jones Lang LaSalle	Denver, Colorado	LEED-EB 2.0 Gold	2009

building type, ownership, and location on the certification. In this chapter, you'll find eight case studies of noteworthy LEED-EB certified projects. Table 13.5 lists some of the characteristics of the case studies presented in this chapter. We've separated out these case studies from the others in this book, so readers could study them as a group.

Sebesta Blomberg Headquarters, Roseville, Minnesota

The Sebesta Blomberg headquarters, a one-story 49,000-square-foot office building, earned 41 LEED-EB v. 2.0 points to achieve Silver rating in 2007 (Fig. 13.1). As an engineering firm with more than 30 LEED Accredited Professionals on staff, Sebesta Blomberg directed a team of internal engineers as well as facility staff to pursue LEED certification for its corporate headquarters in Roseville, Minnesota. President Tony

Figure 13.1 The headquarters of nationally acclaimed Sebesta Blomberg engineering firm achieved a Silver LEED rating in 2007. *Courtesy of Sebesta Blomberg & Associates.*

Litton said that one of the significant reasons for pursuing the LEED-EB certification was to provide that service to their clients.*

> The business case is really multifold, and it surrounds environmental comfort inside the space. A lot of things that we did were invisible such as the recycling program, use of paper and things like that. The normal occupants of the building don't see that. It provides the benefit, and for us, being in a position of providing this service, it put us in the owner's perspective in understanding some of the challenges that our clients see when they're going through the LEED process. It made us better able and capable of delivering the service.

Built in 1999, the Sebesta Blomberg headquarters already had a number of LEED-compliant features such as stormwater management, water-efficient fixtures, daylight and views to the outdoors, and building systems controllability.† Building upon its advantaged starting point in the LEED certification process, the project team sought "to identify low- to no-cost improvements and implement sustainable operation strategies into everyday activities."‡

*Interview with Tony Litton and other members of the Sebesta Blomberg project team, March 2009.
†www.sebesta.com/downloads/SB%20LEED.pdf, retrieved May 29, 2009.
‡Ibid.

Owing to the facility's open layout and the high degree of exterior glazing, the project team anticipated that the building would easily obtain the LEED daylighting credits. Although the Sebesta Blomberg headquarters did receive all of the available points for daylighting; the team was surprised to find that sufficient daylighting (minimum 2 percent daylighting factor) was limited to 52 percent of the building's net occupied area that received the minimum 2 percent daylight factor.

The project team also pursued the following green solutions to achieve the LEED Silver rating:

- *Reduction in unsustainable printing and paper use.* Reevaluated paper and printing use associated with the firm's routine engineering functions and contracted a sustainable supplier to reduce paper and printing related waste and consumption without incurring significant costs.
- *Facility maintenance and cleaning.* Earned five LEED points by collaborating with Marsden Building Maintenance, the building's external housekeeping service to learn and implement green cleaning practices.
- *Waste stream audit.* Conducted audit of the ongoing consumables waste stream to gauge the baseline amount of waste leaving the facility and used this data to inform ongoing recycling and waste reduction efforts.
- *Emission reduction and documentation.* Employed Leonardo Academy's Cleaner and Greener Program to document the building's reduced CO_2, SO_2, NO_x, PM_{10}, and mercury emissions as a result of its energy efficiency upgrades.
- *Green education.* Actively engaged a range of stakeholders, including the firm's facility management, suppliers, employees, and surrounding community to learn about the building's new and ongoing green strategies. A building tour created for the certification has been used as the basis for several professional society meetings.

The LEED certification of the Sebesta Blomberg headquarters is a classic example of a professional service company committing itself to the same business practices and solutions it advocates for its clients—effectively measuring, managing, and communicating sustainability progress while remaining committed to economic operations.

The Crestwood Building, Atlanta Metro Area, Georgia*

Georgia's first LEED-EB certified office building, the Crestwood Building is a five-story, 93,554-square-foot building set on a 5.6-acre lot outside of Atlanta. The Crestwood demonstrates how incremental and modest green building improvements can culminate in LEED certification and market differentiation.

Originally built in 1986, the Crestwood was purchased by Melaver, Inc. in 1998 for $9.4 million (Fig. 13.2). Based in Savannah, Georgia, Melaver holds a predominantly

*Source: *The Green Building Bottom Line*, pp. 167–193.

Figure 13.2 **According to Melaver's director of portfolio management, Scott Soksansky, "we are always on the lookout for small improvements and adjustments to our day-to-day routines that can save energy, reduce waste, and improve tenant comfort, satisfaction, and retention."*** *Courtesy of Melaver, Inc.*

green building real estate portfolio and in 2002 committed itself to pursue LEED certification on all its new developments. (Melaver created the first LEED-certified shopping center in the country in 2006, Abercorn Common in Savannah.)† The Crestwood building, however, presented Melaver with its first viable opportunity to green an existing building.

Before initiating the LEED-EB certification process for the Crestwood in 2004, Melaver first assessed the building's ENERGY STAR rating. The Crestwood required window film coating and upgrades to its lighting and HVAC systems to achieve an ENERGY STAR label. Crestwood's ENERGY STAR upgrades cost $63,000, representing slightly under half of the total expenditures required to achieve LEED-EB certification.‡

Once the ENERGY STAR efficiency measures were addressed, Melaver's internal project team evaluated each LEED-EB credit category to identify the most feasible,

*Ibid., p. 193.

†www.abercorncommon.com, retrieved May 29, 2009.

‡*The Green Building Bottom Line*, op. cit., p. 178.

cost-effective green building improvements for the Crestwood. The following strategies illustrate the various ways in which the Crestwood project team amassed 35 total LEED-EB 2.0 points to achieve basic Certified status:

- *Green site.* Contracted with a landscape architect and contractor to capitalize on the property's existing tree shade cover by strategically placing additional canopy trees to meet the 30 percent shading requirement for heat island reduction.
- *HVAC improvements.* Rebuilt the building's chiller for $15,420 to improve indoor air quality and to reduce energy use.
- *Reduced water use.* Renovated the building's six bathrooms and reconfigured the landscape watering to achieve a 30 percent reduction in overall water consumption.*
- *Reduced electricity use.* Utility costs increased by 25 to 40 percent in Atlanta after Hurricane Katrina in 2005. The Crestwood absorbed these increases and actually saved an estimated $26,000 from 2005 to 2008 as a result of installing new high-efficiency fluorescent lamps, photocell-based light sensors, motion-detector lighting, and energy-efficient vending machines.
- *Comprehensive recycling program.* Supplied recycling bins, collected and stored recyclable metals, plastics, glass, batteries, paper, cell phones, toner cartridges, and light bulbs until pickup for an annual recycling cost of $2860.
- *Low-impact cleaning policy.* Applied Melaver's internal *Mark of a Difference: Interior Care* standards that include daytime janitorial cleaning and the use of sustainable, Green Seal-approved cleaning products.
- *Entryway systems.* Installed a durable metal entry grate/footmat at the building's front entry to reduce the debris, dirt, and pollutants occupants track into the building.

All in all, Melaver invested $137,121 to obtain an ENERGY STAR label and LEED-EB Certified rating for the Crestwood. Over the investment period, Melaver has experienced an above-market occupancy rate of 92 percent, compared to 81 percent within its submarket. Although the Crestwood's higher occupancy cannot be attributed solely to its green upgrades, Melaver estimates the building's higher occupancy increases the property's net income by $109,000 (and probably adds about $2 million to its market value).

Alliance for a Sustainable Colorado, Denver, Colorado†

In 2004, the Alliance for Sustainable Colorado purchased a 38,600-square-foot warehouse in Lower Downtown Denver ("LoDo"), originally built in 1908, and renovated it to create the Alliance Center, a hub for a growing network of sustainability organizations

*www.melaver.com, retrieved May 29, 2009.

†Aaron Nelson, "Alliance Center: A Model of Collaboration," information provided to the author in March 2009.

Figure 13.3 **The Alliance for a Sustainable Colorado took a 1908 brick warehouse and upgraded it into a LEED-EB Gold center housing 30 nonprofit tenants, all working toward sustainability in Colorado. Lobby displays informing visitors of the project's LEED certification measures help to advance the organization's educational mission.** *Courtesy of Alliance for Sustainable Colorado.*

in Colorado (Fig. 13.3). Currently the Alliance Center houses 30 nonprofit tenants, including organizations which focus on renewable energy, resource conservation, government ethics, and advocacy for environmental protection. Inherent in the building's transformation was the goal of creating a leading example of a green office building. The Alliance completed building renovations in two phases; the first phase, concluded in early 2005, created offices and meeting rooms. The second phase, concluded in early 2006, upgraded resource-efficient equipment and fixtures.

Home to about 110 occupants, in June 2006 the Alliance Center received LEED-EB Gold certification and in August 2006 earned LEED-CI Silver certification, becoming the first building in the world to achieve LEED certifications in these two categories. The Center also earned the ENERGY STAR Leader status from the U.S. EPA.

The Alliance installed educational signs throughout the building. These signs facilitate self-guided tours of the Alliance Center, complemented by guided tours given by Alliance staff and a virtual tour provided through the organization's Web site.

After the implementation of the LEED-EB and LEED-CI project measures, the Alliance Center exhibited significant utility use and cost reductions. The Alliance removed all of the building's preexisting plumbing fixtures and installed waterfree urinals, low-flow toilets, showerheads, and faucets. The Alliance Center estimates water cost savings are more than $4500 per year. The total plumbing fixture retrofits cost

$22,000, which translates into a payback period of less than five years. These plumbing modifications coupled with a water conservation initiative among the building occupants immediately reduced water consumption by 84 percent compared to the previous month. Despite water rates increasing at a rate of 7 percent since 2005, the Alliance Center still saves nearly 70 percent in annual water use costs and currently uses over 90 percent less water than it did in 2005. Additionally, there was a 10 percent reduction in energy use just from LEED-required commissioning and the replacement of older pneumatic HVAC controls with digital HVAC controls.

The project also achieved 40 percent waste diversion from landfills through increased recycling during construction and the implementation of a single-stream recycling program. Similarly, all of the furniture in the building is either reused or Greenguard-certified. And after the renovation, more than 50 percent of occupants now have access to daylit workspaces.

The Alliance project team highlighted the value of the LEED standards during the renovation of a 100-year-old historic building: "*LEED gave us benchmarks by which to judge upgrades to our building and the flexibility to decide whether the best plan was complete replacement, or creative modifications to what already existed.*" More than 55 percent of the existing building was reused during the Alliance Center renovation, demonstrating that the most sustainable building is often the one you reuse and don't build from scratch.

Byron G. Rogers U.S. Courthouse, Denver, Colorado*

In 2006, the Byron G. Rogers U.S. Courthouse became the first federal building to be awarded a LEED-EB Gold rating (Fig. 13.4). Located in downtown Denver, the courthouse is owned and operated by the General Services Administration (GSA) and consists of 260,000 square feet of courtrooms and office space plus two levels of underground parking. The courthouse had been used continuously, without any significant upgrades, since its original construction in 1965.

In 2002, the GSA began a four-year period of redesign and renovation. The $46 million renovation project was managed by the GSA and Bennett Wagner & Grody Architects who were tasked with modernizing the courthouse without diminishing the building's architecture, history, and future eligibility for inclusion in the National Register of Historic Places.

The project's key sustainable strategies—building reuse and preservation of character-defining features—were complicated by the fact that the building had asbestos- and lead-based paints throughout the facility. Both the asbestos- and lead-based paints required extensive abatement programs. The renovation specifications

*"Byron G. Rogers, U.S. Courthouse Renovation, Denver, Colorado." GSA Environmental Award 2007, GSA publication.

Figure 13.4 In the early 2000s, Denver's Byron G. Rogers U.S. Courthouse combined a major building renovation with green operating practices to earn a LEED-EB v.1.0 Gold rating. *Courtesy of GSA's Public Buildings Service.*

also clearly defined the cleaning and reuse of much of the original interior wood wall paneling, judges' benches, marble panels, brass elevator doors and frames, and a range of decorative metal, wood, and plaster reveals and moldings. The courthouse's site plan, shell, and structure were also retained.

Don Horn, director of the Sustainability Program of GSA's Public Buildings Service, responsible for the overall sustainability of the GSA's building stock, acknowledges the added creditability of a LEED-certified efficiency improvement project. According to Horn, the Byron G. Rogers U.S. Courthouse renovation shows:

> The LEED rating system provides documentation that says you are actually doing what you say you are doing. The value of LEED certification (that we don't otherwise have) is the tracking and reporting mechanism."*

The Byron G. Rogers U.S. Courthouse renovation earned a total of 44 points under the LEED-EB Pilot Project certification system, including meeting 13 prerequisites. Achievements in each LEED category included the following:

- Sustainable Sites: 11 out of 16 possible
- Water Efficiency: 2 points out of 5 possible

*Interview with Don Horn, March 2009.

- Energy and Atmosphere: 15 points out of 22 possible
- Materials and Resources: 2 points out of 10 possible
- Indoor Environmental Quality: 10 points out of 18 possible
- Innovation and Design Process: 4 points out of 5 possible

The courthouse project team used the following green operating strategies to achieve its LEED certification:

- *Planning.* Held a sustainable design charrette to discuss design strategies and integrate design concepts. Energy modeling was also performed throughout the design process to optimize and select the building's mechanical, electrical, and architectural systems.
- *Water conservation.* Reduced water use more than 20 percent compared with the LEED-EB baseline, by installing new low-flow plumbing fixtures.
- *Recycling improvements and savings.* Increased recycling by 20 percent between 2001 and 2006, which saved the GSA $4718 in 2006 on waste disposal cost.
- *Green housekeeping policy.* Used disposable paper products with 100 percent recycled content as well as a low-environmental-impact pest management policy.
- *Sustainable site features.* Increased open space with more than 50 percent of the site landscaping using native or adapted vegetation.
- *Sustainable wood products.* Nearly all the new wood installed during the renovation was FSC-certified as sustainably harvested wood.
- *Daylighting.* To promote daylighting, the project added new exterior windows, redesigned the building lobby with floor to ceiling glazing, and inserted clerestory windows into interior walls to allow natural and "borrowed" light to penetrate the interior office spaces.
- *Improved air quality.* Installed separate exhausts for the copy rooms and janitor closets and adopted green cleaning practices, including using environmentally friendly cleaning products.
- *Renewable energy use.* The courthouse is powered with 100 percent purchased wind power.

City Hall, San Jose, California*

In 2009, San Jose City Hall earned the distinction of being the first city hall in the United States to obtain a LEED-EB Platinum rating (Fig. 13.5). The 530,000-square-foot City Hall, owned and operated by the City of San Jose, contributes to the city's ambitious *Green Vision* sustainability plan. Among the 10 goals defined in San Jose's *Green Vision*, the city challenged itself to build or retrofit 50 million square feet of building space citywide to green building space by 2022, achieve zero waste to landfill,

---

**San José City Hall is First City Hall in the Nation to Achieve LEED Platinum Rating*, City of San Jose News Release, March 9, 2009. All quotes from that release.

Figure 13.5 As a participant in the USGBC's Portfolio Program, the City of San Jose has committed to certifying the operation and maintenance of at least 30 additional existing buildings, with 4 planned for certification in 2009.
Photo by Paul Marino.

sharply reduce energy use, and reuse 100 percent of the city's wastewater. San Jose Mayor Chuck Reed noted the measurable significance the City Hall certification had toward the city's green objectives. "Certifying City Hall dramatically increases our municipal square footage of green building space and brings our new total to over 700,000 square feet."

In announcing the certification award, the mayor also emphasized the business case for the City Hall renovations: "In difficult economic times, improving building performance helps the city's bottom line. By being energy-efficient and implementing other green practices, San Jose's City Hall will yield annual savings of more than $30,000 in energy costs alone."

A cross-functional team of employees from San Jose's Departments of General Services, Finance, Environmental Services, Information Technology, and Parks, Recreation, and Neighborhood Services developed and implemented the following green strategies for the project:

- *Energy conservation.* Installed high-efficiency lighting and windows and introduced additional daylighting features, which resulted in an ENERGY STAR label and a building that is more energy efficient than 93 percent of comparable facilities.
- *Reduced water use.* Reduced City Hall's water use by 82 percent from a combination of adopting water conserving practices, installing ultra-low-flow fixtures, and using gray water for landscape irrigation.

- *Waste reduction and diversion.* Diverted over 90 percent of the City Hall's surplus furniture and office equipment, electronics, and construction waste from the landfill through finding and adopting new recycling solutions.
- *Environmental purchasing.* Applied the city's Environmentally Preferable Purchasing Policy (EP3), which promotes the purchase of green cleaning products, office supplies, furniture, computers, and building materials.
- *Indoor air quality.* Optimized indoor air quality by using high-efficiency filters in the HVAC systems, along with green cleaning products and practices.

The San Jose City Hall LEED Platinum certification serves to emphasize the city's commitment to citywide sustainability practices. Furthermore, the city's approach to sustainability illustrates the recent momentum behind greening existing buildings as well as the long-term commitment many municipalities are starting to make to LEED certification.

Bell Trinity Square, Toronto, Ontario*

In 2009, Bell Trinity Square (Fig. 13.6) became the first LEED-EB 2.0 Gold certified building in Toronto and the second in Canada. With nearly 1 million square feet of gross leasable area and comprising two towers, Bell Trinity Square is a Class A office property located at 483 Bay Street, adjacent to Toronto City Hall. This Toronto office building is owned by Germany-based, Union Investment Real Estate AG, and is managed by Northam Realty Advisors Limited, Toronto.

The LEED project team was led by Northam and consisted primarily of the building's management and facilities personnel. Beginning in 2006, the project team studied and selected building upgrades and sustainable practices that would enable the building to achieve LEED certification. Bell Trinity Square's green solutions included:

- *Building and systems modeling.* Prior to pursuing sustainability opportunities, the project team used a standard energy modeling tool, DOE-2, to estimate performance benefits of system upgrades.
- *Reduce energy use.* Installed high-efficiency chillers, boilers, and lighting systems to reduce energy use and the building's carbon footprint.
- *Load shifting of electrical demand.* Chilled water is produced at night when off-peak electricity rates are one-third of peak daytime rates. This chilled water is then stored in the building's 1.2-million-gallon (4.5-million-liter) thermal energy storage facility.
- *Motion detectors.* Maximized the benefit of lighting and HVAC system controls with motion detectors for more efficient lighting and heating and cooling energy use.
- *Enhancing indoor air and environmental quality.* Improved the building's volume of fresh air supply and enhanced environmental filtration systems.

**Bell Trinity Square sets the gold standard for green*, Northam Realty Advisors Limited News Release, April 22, 2009.

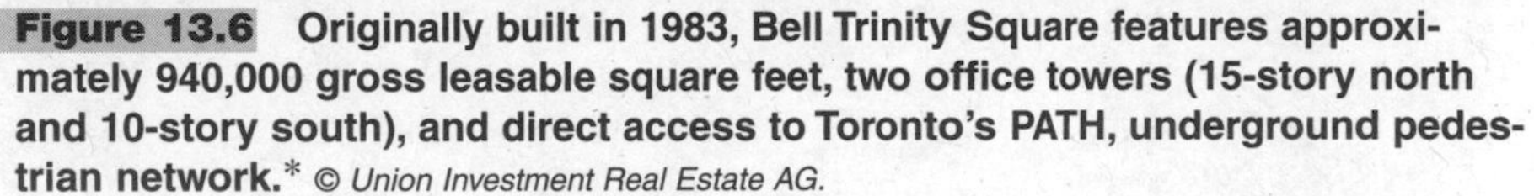

Figure 13.6 **Originally built in 1983, Bell Trinity Square features approximately 940,000 gross leasable square feet, two office towers (15-story north and 10-story south), and direct access to Toronto's PATH, underground pedestrian network.*** *© Union Investment Real Estate AG.*

The Bell Trinity renovation earned a total of 49 points under the LEED-EB 2.0 certification system in addition to meeting the system's 13 prerequisites.† Achievements in each LEED category included the following:

- Sustainable Sites: 6 out of 14 possible
- Water Efficiency: 2 points out of 5 possible
- Energy and Atmosphere: 13 points out of 23 possible
- Materials and Resources: 11 points out of 16 possible
- Indoor Environmental Quality: 12 points out of 22 possible
- Innovation and Design Process: 5 points out of 5 possible

*http://www.northamrealty.com/assetPortfolio/BellTS.shtml, accessed July 24, 2009.

†Bell Trinity Square, LEED-EB 2.0 Application Review, March 25, 2009.

Credit highlights included a 17 percent reduction in annual carbon-dioxide emissions, the diversion away from landfill of 80 percent of all solid waste generated during the performance period, the implementation of certified green practices, and water reductions of 24-million gallons (92-million liters) annually.

Northam Realty Advisor's general manager at Bell Trinity Square, William Braun, highlighted the environmental and economic benefits of LEED certification when he commented that, "Being environmentally sustainable is responsible and it makes good business sense—long-term cost savings are benefiting existing tenants and we attract new tenants by demonstrating that we are industry leaders."

One Potomac Yard, Arlington, Virginia*

One Potomac Yard illustrates the relatively quick progression new buildings can experience from LEED certification for New Construction (LEED-NC) to LEED-EB (Fig. 13.7). The 654,000-square-foot office complex features two 12-story buildings and is owned by JP Morgan Asset Management and managed by Jones Lang LaSalle.

Figure 13.7 One Potomac Yard illustrates effective private-public sector collaboration as the developer, property manager, and owner worked with the lessee (GSA) and the lead tenant (EPA) to accommodate the federal government's sustainability requirements. *Photo by Daniel Hart, Courtesy of the U.S. EPA.*

**Potomac Yards 1 and 2/EPA*, U.S. General Service Administration Case Study, from http://www.gsa.gov/Portal/gsa/ep/contentView.do?contentType=GSA_BASIC&contentId=21865, accessed June 3, 2009.

In 2004, Potomac Yard was initially designed as a speculative office building without any green specifications. The original developer, Crescent Resources LLC, effectively repositioned Potomac Yard as a LEED-certified building in order to meet the lease requirements of its anchor tenant, the U.S. Environmental Protection Agency (EPA).* After achieving a LEED-NC Gold certification in 2006, Potomac Yard received a LEED-EB Gold certification in 2008.

The project team included representatives from Crescent Resources and the EPA as well as an environmental building consultant and a commissioning authority. This project incorporated the following green strategies, many of which originated in the LEED-NC certified project and also met LEED-EB version 2.0 criteria:

- *Roofing materials.* Used highly reflective and ENERGY STAR-compliant roof materials to reduce the building's solar heat gain and to decrease related cooling demand and energy use.
- *Green roofing.* An elevated patio with benches made of recycled plastic lumber and a 1711-square-foot green roof with sedum plants connects the two office towers.
- *Brownfield site remediation.* Prior to construction, the developer removed soil contaminants, including cinder ballast, arsenic, and benzo(A)pyrene.
- *On-site stormwater treatment.* Below grade sand filters treat stormwater runoff from each building before it flows into the Potomac River. The treatment system is expected to reduce total suspended solids (silt) by 80 percent and total phosphorous by 40 percent.
- *Environmentally responsible commuting.* The project provides on-site bicycle parking for 53 bikes, shower facilities, and close proximity to existing Metrorail lines, Metro buses, and EPA shuttle-bus routes.
- *Low-flow plumbing fixtures.* The project reduced water use by 41 percent through installation of dual-flush toilets, ultra-low-flush urinals, low-flow showerheads, and ultra-low-flow lavatory faucets.
- *No permanent irrigation.* The project landscaping has drought-resistant and local plants that can survive without a permanent irrigation system.
- *ENERGY STAR-efficient lighting strategies.* Installed glass panel insets in the systems, furniture, glass doors, and sidelights for conference rooms and offices that extend natural daylighting well into the occupied space.
- *Green power.* The building purchased green power to meet 68 percent of its electrical energy needs for two years.
- *Sustainably harvested wood.* FSC-certified wood-based materials constituted 83 percent of the value of all wood used in the building.
- *Low-VOC adhesives, paints, sealants, and caulks.* Tenant and public areas used low-volatile-organic-compound (VOC) materials and finishes to ensure high indoor air quality (IAQ).
- *Green cleaning and pest management.* The building managers adopted sustainable best practices for cleaning, pest management, and landscaping.

**Sustainable Facilities and EPA: One and Two Potomac Yard*, U.S. Environmental Protection Agency-Fact Sheet, from http://www.epa.gov/greeningepa/documents/py_factsht_508.pdf, accessed May 31, 2009.

- *Sustainable workstation furniture.* The building management procured furniture with 35 to 40 percent recycled material content as required under the federal government's Comprehensive Procurement Guidelines (CPG) program. The furniture was also Greenguard certified for air emissions.
- *Local materials.* The project received an innovation credit for exemplary performance in sourcing local materials, because 62.8 percent of the total materials by cost were manufactured within 500 miles.
- *Occupant education program.* Another innovation credit was achieved because the building used signage to inform occupants and visitors about the sustainable features and operations of the facility.

17th Street Plaza, Denver, Colorado*

In 2000, JP Morgan Asset Management purchased 17th Street Plaza, a 32-story Class A office building in Denver's central business district (Fig. 13.8). Built in 1982, this 666,000-square-foot office tower was placed under the management of Jones Lang LaSalle (JLL).† Chief engineer Curt Godes led an ongoing, eight-year effort to improve the building's performance and sustainability. Between 2000 and 2008, the property team spent $1,222,084 in capital projects, resulting in annual average electricity savings of $202,000 (from saving 4 million kilowatt-hours) and water savings of $15,630. When JP Morgan encouraged Godes' team to pursue LEED-EB certification in 2007, minimal additional work was required to achieve the building's LEED Gold rating in 2009.

Prior to pursuing LEED certification, the 17th Street Plaza building managers employed the following performance and sustainability strategies:

- *Utility rebate program.* Used the demand side management (DSM) utility rebate program developed by one of the building's lead tenant, Xcel Energy, to help pay for demand-reduction projects.
- *ENERGY STAR rating.* In 2001, the building earned its ENERGY STAR label with a score of 77 and has since improved its score. In 2008, the building scored 94 to rank among the top 6 percent of all office buildings in energy efficiency.
- *Low-mercury lighting.* JLL replaced standard fluorescent lamps with comparably priced low-mercury lamps that met the LEED-EB criteria.
- *Recycling program.* Established a recycling program that redirected 219 tons of waste from landfills and induced the building's largest tenant, Xcel Energy, to implement the same program throughout its 350,000-square-foot Denver portfolio.

Having already implemented many sustainable building practices and achieved an ENERGY STAR rating, the building team used the LEED certification process primarily

---

**Building Sustainable Value at 17th Street Plaza*, Jones Lang LaSalle Case Study, from http://newyork.uli.org/Events/Past%20Events/Content/~/media/DC/New%20York/NY%20Docs%202/YLG%20Retrofit%20Handout.ashx, accessed May 31, 2009.

†*Jones Lang LaSalle, 17th Street Plaza Property Details*, from http://www.costar.com/costarconnect/MasterPage/main.aspx?SiteID=20865&Checksum=87669&Demo=0&RtnURL&LogRedirect=1

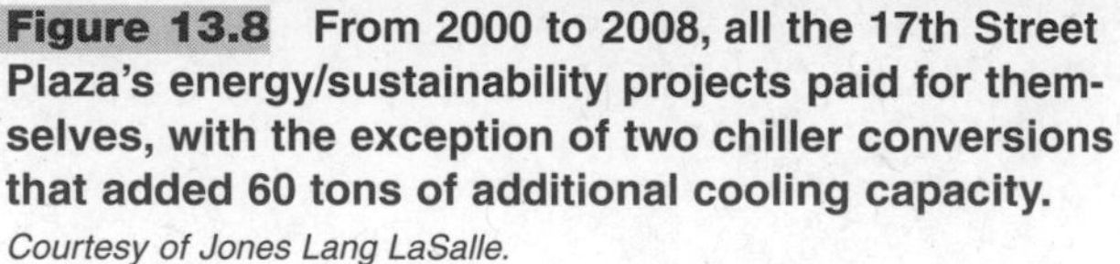

Figure 13.8 From 2000 to 2008, all the 17th Street Plaza's energy/sustainability projects paid for themselves, with the exception of two chiller conversions that added 60 tons of additional cooling capacity.
Courtesy of Jones Lang LaSalle.

to communicate with tenants and verify the building's market leadership in performance and sustainability. To achieve LEED certification, the project team's modest building improvements included the following water conservation upgrades:

- *Toilet and urinal flush valve kits.* Spent $1252 to install low-flow flush valve kits on toilets and urinals, resulting in annual water savings of $5336.
- *Lavatory aerators.* Installed 0.5-gallon-per-minute aerators on restroom faucets and 1.5-gallon-per-minute aerators on kitchen faucets at a cost of $300.
- *Weather satellite-controlled landscape irrigation.* Installed Weather TRAK-ET irrigation controllers for a cost $3800. The system has reduced irrigation water costs by $5894 in the first year of operation and is estimated to reduce annual irrigation water usage by 20 percent.

At 17th Street Plaza, JLL proved that with a reasonably well-designed building, the distinction between cost-savings, tenant-focused building management and LEED-EB certification for existing buildings should be minimal. As JLL stated, “While most projects [at 17th Street Plaza] were building improvement efforts rather than sustainability efforts, in many cases increasing efficiency, choosing “greener” options and saving money went hand in hand.”

I

CERTIFIED LEED-EB PROJECTS (ALL VERSIONS) (LIST CURRENT AS OF MAY 28, 2009)*

PROJECT NAME	CERTIFICATION LEVEL	VERSION	CITY	STATE
San Jose City Hall	Platinum	EBOM	San Jose	CA
FBI Regional Office	Platinum	EBOM	Chicago	IL
Center Tower	Gold	EBOM	Costa Mesa	CA
Columbia Center	Gold	EBOM	San Diego	CA
Hotel Carlton	Gold	EBOM	San Francisco	CA
Park Tower	Gold	EBOM	Costa Mesa	CA
Plaza Tower	Gold	EBOM	Costa Mesa	CA
Watt Plaza	Gold	EBOM	Los Angeles	CA
17th Street Plaza	Gold	EBOM	Denver	CO
Hurt Building	Gold	EBOM	Atlanta	GA
Wrigley Global Center—Transwestern	Gold	EBOM	Chicago	IL
One Boston Place	Gold	EBOM	Boston	MA
CTA Billings Office	Gold	EBOM	Billings	MT
Bank of America Plaza	Gold	EBOM	Charlotte	NC
Symantec Springfield	Gold	EBOM	Springfield	OR

(*Continued*)

*Data from USGBC Certified Project List, www.usgbc.org/leed, accessed May 28, 2009.

PROJECT NAME	CERTIFICATION LEVEL	VERSION	CITY	STATE
The Standard: Plaza Building	Gold	EBOM	Portland	OR
First City Tower	Gold	EBOM	Houston	TX
5055 Wilshire	Silver	EBOM	Los Angeles	CA
Northrop Grumman Plaza	Silver	EBOM	Los Angeles	CA
555 17th Street	Silver	EBOM	Denver	CO
The Quadrant Building	Silver	EBOM	Greenwood Village	CO
245 Summer Street	Silver	EBOM	Boston	MA
300 Puritan Way	Silver	EBOM	Marlboro	MA
400 Puritan Way	Silver	EBOM	Marlboro	MA
One Financial	Silver	EBOM	Boston	MA
Capital Gateway 1&2 —Transwestern	Silver	EBOM	Bethesda	MD
Accenture Tower	Silver	EBOM	Minneapolis	MN
520 Madison Avenue	Silver	EBOM	New York	NY
MetLife	Silver	EBOM	Long Island City	NY
1800 West Loop South —Transwestern	Silver	EBOM	Houston	TX
717 Texas—Hines	Silver	EBOM	Houston	TX
MacArthur Plaza —Transwestern	Silver	EBOM	Irving	TX
101 California	Certified	EBOM	San Francisco	CA
Office Building B264	Certified	EBOM	Livermore	CA
Colorado State Capitol	Certified	EBOM	Denver	CO
The Miami Heat Goes Green	Certified	EBOM	Miami	FL
Pershing Point Plaza	Certified	EBOM	Atlanta	GA
The AO Project	Certified	EBOM	Atlanta	GA
Long Lake Crossing —Transwestern	Certified	EBOM	Troy	MI
Butler Square	Certified	EBOM	Minneapolis	MN
45 Earhart Drive	Certified	EBOM	Buffalo	NY
Lincoln Center II & III —Transwestern	Certified	EBOM	West Allis	WI
One Washingtonian Center—Transwestern	Certified	EBOM	Gaithersburg	MD

PROJECT NAME	CERTIFICATION LEVEL	VERSION	CITY	STATE
William J. Clinton Presidential Center	Platinum	2.0	Little Rock	AR
601 Townsend Adobe Systems, Inc.	Platinum	2.0	San Francisco	CA
Adobe Systems Inc. Almaden Tower	Platinum	2.0	San Jose	CA
Adobe Systems Inc. East Tower	Platinum	2.0	San Jose	CA
Adobe Tower West	Platinum	2.0	San Jose	CA
CA Department of Education Building	Platinum	2.0	Sacramento	CA
SMUD Customer Service Center	Platinum	2.0	Sacramento	CA
Vision Service Plan (VSP) HQ1	Platinum	2.0	Rancho Cordova	CA
McDonald's Corporation —World HQ	Platinum	2.0	Oakbrook	IL
Discovery Communications Global HQ	Platinum	2.0	Silver Spring	MD
Cascade Engineering—Learning Community	Platinum	2.0	Grand Rapids	MI
E & E Corporate Headquarters	Platinum	2.0	Lancaster	NY
Armstrong World Industries, Corporate HQ	Platinum	2.0	Lancaster	PA
245 Market	Gold	2.0	San Francisco	CA
AT&T Center Los Angeles	Gold	2.0	Los Angeles	CA
CA Dept. of Public Health/ Healthcare	Gold	2.0	West Sacramento	CA
Hawthorne Plaza	Gold	2.0	San Francisco	CA
RRM Corporate Headquarters	Gold	2.0	San Luis Obispo	CA
Synopsys Mountain View Campus	Gold	2.0	Mountain View	CA
Alliance Center	Gold	2.0	Denver	CO
Boulder Courthouse Annex	Gold	2.0	Boulder	CO
Porter Industries Building	Gold	2.0	Loveland	CO
The Wellington E. Webb Building	Gold	2.0	Denver	CO
USAA South East Regional Office	Gold	2.0	Tampa	FL
ABN Amro Plaza	Gold	2.0	Chicago	IL
Chicago Transit Authority Headquarters	Gold	2.0	Chicago	IL
Food Bank of Western Massachusetts	Gold	2.0	Hatfield	MA

(*Continued*)

PROJECT NAME	CERTIFICATION LEVEL	VERSION	CITY	STATE
NSTAR Corporate Headquarters	Gold	2.0	Westwood	MA
Sasaki Associates, Inc.	Gold	2.0	Watertown	MA
College of Health & Human Services	Gold	2.0	Kalamazoo	MI
LaSalle Bank Financial Center	Gold	2.0	Troy	MI
Allan L. Schuman Corporate Campus	Gold	2.0	Eagan	MN
Ameriprise Client Service Center	Gold	2.0	Minneapolis	MN
Crescent Ridge Corporate Center I	Gold	2.0	Minnetonka	MN
Greening the NYS Governor's Mansion	Gold	2.0	Albany	NY
New York Power Authority White	Gold	2.0	White Plains	NY
PepsiCo Valhalla R&D Building	Gold	2.0	Valhalla	NY
200 Market Place	Gold	2.0	Portland	OR
Vernier Software & Technology	Gold	2.0	Beaverton	OR
Carriage House Children's Center	Gold	2.0	Pittsburgh	PA
Caterpillar Financial Headquarters	Gold	2.0	Nashville	TN
One Potomac Yard	Gold	2.0	Arlington	VA
Starbucks Center	Gold	2.0	Seattle	WA
APS Flagstaff Administrative Building	Silver	2.0	Flagstaff	AZ
APS Wickenburg Service Center	Silver	2.0	Wickenburg	AZ
Bentley Prince Street	Silver	2.0	City of Industry	CA
Engineering 2—Univ. of Calif.	Silver	2.0	Santa Cruz	CA
Faculty Offices East—Cal Poly State Univ.	Silver	2.0	San Luis Obispo	CA
Girvetz Hall—Univ. of Calif.	Silver	2.0	Santa Barbara	CA
MGM Tower	Silver	2.0	Los Angeles	CA
Pleasanton Corporate Commons	Silver	2.0	Pleasanton	CA
The Ambrose	Silver	2.0	Santa Monica	CA
Office of the President—Univ. of Calif.	Silver	2.0	Oakland	CA
UCSB Recreational Center	Silver	2.0	Santa Barbara	CA
4600 S. Syracuse	Silver	2.0	Denver	CO
Golden Hill Office Centre	Silver	2.0	Lakewood	CO
University of Colorado Memorial Center A	Silver	2.0	Boulder	CO

PROJECT NAME	CERTIFICATION LEVEL	VERSION	CITY	STATE
222 Merchandise Mart	Silver	2.0	Chicago	IL
One North Wacker	Silver	2.0	Chicago	IL
PepsiCo Chicago Headquarters	Silver	2.0	Chicago	IL
One Beacon Street	Silver	2.0	Boston	MA
Ameriprise Financial Center	Silver	2.0	Minneapolis	MN
Sebesta Blomberg & Associates Corporate	Silver	2.0	Roseville	MN
135 West 50th Street	Silver	2.0	New York	NY
320 Park Avenue	Silver	2.0	New York	NY
Owens Corning World Headquarters	Silver	2.0	Toledo	OH
Avalon Hotel and Spa	Silver	2.0	Portland	OR
Liberty Centre	Silver	2.0	Portland	OR
Oregon Convention Center	Silver	2.0	Portland	OR
Bridgestone Firestone Warren Green Bldg.	Silver	2.0	Morrison	TN
2000 Edmund Halley Drive	Silver	2.0	Reston	VA
2002 Edmund Halley Drive	Silver	2.0	Reston	VA
Nulhegan Administration/Visitor Contact	Silver	2.0	Brunswick	VT
16 W Harrison	Silver	2.0	Seattle	WA
901 5th Avenue	Silver	2.0	Seattle	WA
Athena Building	Silver	2.0	Madison	WI
John J. Flynn Elementary School	Silver	2.0	Eau Claire	WI
Monona Terrace Community	Silver	2.0	Madison	WI
100 Pine Street	Certified	2.0	San Francisco	CA
Los Angeles Convention Center	Certified	2.0	Los Angeles	CA
Orchard Hotel	Certified	2.0	San Francisco	CA
The Getty Center	Certified	2.0	Los Angeles	CA
Colorado Governor's Residence	Certified	2.0	Denver	CO
Compliance Partners/Seven Generations	Certified	2.0	Fort Collins	CO
Judicial Building & Colorado History Museum	Certified	2.0	Denver	CO
M-E Engineers Office Building	Certified	2.0	Wheat Ridge	CO
State of Colorado—Human Services Building	Certified	2.0	Denver	CO

(*Continued*)

PROJECT NAME	CERTIFICATION LEVEL	VERSION	CITY	STATE
State of Colorado—State Services Building	Certified	2.0	Denver	CO
Stop & Shop (51 stores)	Certified	2.0	Various	CT
				MA
				NH
				NJ
				NY
				RI
The Millennium Building	Certified	2.0	Washington	DC
Crestwood Building	Certified	2.0	Duluth	GA
Securian	Certified	2.0	St. Paul	MN
Janssen Pharmaceutica Recertification	Certified	2.0	Titusville	NJ
Merrill Lynch—Hopewell Campus	Certified	2.0	Pennington	NJ
Tate Snyder Kimsey Architects Design Studio	Certified	2.0	Henderson	NV
New York Mercantile Exchange	Certified	2.0	New York	NY
SAP America Headquarters	Certified	2.0	Newtown Square	PA
1 Independence Pointe	Certified	2.0	Greenville	SC
John J. Duncan Federal Building	Certified	2.0	Knoxville	TN
Rosetta Inpharmatics	Certified	2.0	Seattle	WA
Whatcom County Courthouse	Certified	2.0	Bellingham	WA
Luck K-12 School	Certified	2.0	Luck	WI

INTERNATIONAL PROJECTS	CERTIFICATION LEVEL	VERSION	CITY	COUNTRY
Brandix Casualwear Ltd.	Platinum	2.0	Seeduwa	Sri Lanka
ICAO Building	Gold	2.0	Montreal	Canada
NEG Micon	Gold	2.0	Chennai	India
Bell Trinity Square	Gold	2.0	Toronto	Canada
225 King Street West	Silver	2.0	Toronto	Canada
GE Meadowvale Corporate Headquarters	Certified	2.0	Mississauga	Canada

PROJECT NAME	CERTIFICATION LEVEL	VERSION (PILOT)	CITY	STATE
Joe Serna Jr.—Cal/EPA Headquarters	Platinum	1.0	Sacramento	CA
Swinerton Builders Corporate Headquarters	Gold	1.0	San Francisco	CA
Moss Landing Marine Labs	Gold	1.0	Moss Landing	CA
Byron G. Rogers U.S. Courthouse	Gold	1.0	Denver	CO
Denver Place	Gold	1.0	Denver	CO
Goizueta Business School	Gold	1.0	Atlanta	GA
Len Foote Hike Inn	Gold	1.0	Dawsonville	GA
Karges-Faulconbridge Office Building	Gold	1.0	St. Paul	MN
Nike, Inc., Ken Griffey Jr. Building	Gold	1.0	Beaverton	OR
CCI Center	Gold	1.0	Pittsburgh	PA
Lubin Manufacturing Facility	Gold	1.0	East Greenville	PA
King Street Center	Gold	1.0	Seattle	WA
Brengel Technology Center	Gold	1.0	Milwaukee	WI
JohnsonDiversey Global Headquarters	Gold	1.0	Sturtevant	WI
National Geographic Society Headquarters	Silver	1.0	Washington	DC
Ada County Courthouse and Administration	Silver	1.0	Boise	ID
AJ Martini, Inc.	Silver	1.0	Winchester	MA
Janssen Pharmaceutica Inc.	Silver	1.0	Titusville	NJ
WA Department of Ecology	Silver	1.0	Olympia	WA
Roosevelt Facility	Certified	1.0	Scottsdale	AZ
Duke University—Smith Warehouse	Certified	1.0	Durham	NC
Key Bank—Tiedeman Campus	Certified	1.0	Brooklyn	OH
Oregon Convention Center	Certified	1.0	Portland	OR
Kirksey Corporate Office Building	Certified	1.0	Houston	TX
Microsoft Buildings 30, 31, & 32	Certified	1.0	Redmond	WA

II

ENERGY STAR AND THE EU'S ENERGY PERFORMANCE OF BUILDINGS DIRECTIVE

This appendix presents both the U.S. ENERGY STAR rating system for building energy performance and rating systems based on the European Union's (EU) Energy Performance of Buildings Directive. The voluntary program, ENERGY STAR, is the currently preferred building assessment and labeling program in the United States, but it's likely that aspects of the EU's program will gain recognition and adoption in the United States, as ENERGY STAR labeling becomes more of a mandatory program over the coming five years.

The ENERGY STAR Rating System

LEED-EBOM uses ENERGY STAR as the primary rating tool for building energy performance. To acquire an ENERGY STAR rating, most buildings will use the EPA's (U.S. Environment Protection Agency) Portfolio Manager interactive energy management tool. Using it, building owners, managers, and operators can track and assess energy and water consumption in a secure online environment. This tool helps prioritize investments, identify underperforming buildings, and verify efficiency improvements.* Buildings that achieve a score of at least 75, placing them in the top 25 percentile of all buildings in their category in the EPA's National Energy Performance Rating System, qualify for the ENERGY STAR label.†

The EPA's Portfolio Manager tool allows users to benchmark their building's performance against all commercial buildings in the United States, using data from the U.S. Department of Energy (DOE), Energy Information Administration, and the 2003

---

*http://www.energystar.gov/index.cfm?c=evaluate_performance.bus_portfoliomanager, accessed March 5, 2009.
†http://www.energystar.gov/index.cfm?c=business.bus_bldgs, accessed May 21, 2009.

Commercial Buildings Energy Consumption Survey (CBECS). It also allows comparisons with specific building categories, such as warehouses, schools, and offices. In fact, the EPA uses these comparisons as the basis of its National Energy Performance Rating System.* The next version of CBECS, based on 2007 energy use data, expects to appear sometime in the fall 2009 and will then be used as the baseline for future ENERGY STAR ratings.

Users monitor energy and water use in buildings by creating a private account on ENERGY STAR to track energy consumption and cost data over time. In addition, users can benchmark building energy performance against past performance and against other buildings, assess progress toward energy management goals over time, and identify strategic opportunities for savings.†

PORTFOLIO MANAGER

Portfolio Manager also calculates financial performance, environmental performance, water use, and GHG emissions. The tool allows users to consolidate energy and water data and track key consumption, performance, and cost information portfolio-wide. With this tool, users can:‡

- Track multiple energy and water meters for each facility
- Customize meter names and key information
- Benchmark facilities relative to their past performance
- View percent improvement in weather-normalized source energy
- Monitor energy and water costs
- Share building data with others inside or outside of the organization
- Enter operating characteristics, tailored to each space use category within a building

BUILDING UPGRADE VALUE CALCULATOR§

ENERGY STAR also provides several financial calculators for building owners. For example, consider the Building Upgrade Value Calculator for office buildings, resulting from a partnership between ENERGY STAR and Building Owners and Managers International (BOMA). This calculator was developed as part of BOMA's Energy Efficiency Program (BEEP), a series of courses designed to help commercial real estate professionals improve building energy efficiency performance. The calculator helps property owners assess the financial value of investments in a property's energy performance.

The Building Upgrade Value Calculator estimates the financial impact of proposed investments in energy efficiency for office properties. The calculations are based on data input by the building owners, representing scenarios and conditions at a given property. Required inputs are limited to general characteristics of the building, plus

*http://www.energystar.gov/index.cfm?c=evaluate_performance.bus_portfoliomanager, accessed March 5, 2009.
†Ibid.
‡Ibid.
§http://www.energystar.gov/index.cfm?c=comm_real_estate.building_upgrade_value_calculator

information on the proposed investments in energy efficiency upgrades. The calculator's output includes the following information:

- Net investment
- Reduction in operating expense
- Energy savings
- Return on investment (ROI)
- Internal rate of return (IRR)
- Net present value (NPV)
- Net operating income (NOI)
- Impact on asset value

In addition to the above outputs, the calculator also estimates the impact the proposed changes will have on a property's ENERGY STAR rating. The tool provides two ways to output calculated results: users can save and print a summary of their results, or they can generate a letter that highlights the financial value of the changes, for use as part of a capital investment proposal.

European Union Energy Performance of Buildings Directive (EPBD)

The EU's EPBD is an important step toward labeling all commercial and institutional buildings, something that will definitely appear in the United States within the next few years. It is based on a similar principle to ENERGY STAR: compare building energy use to a national average and give it a grade or a score. While ENERGY STAR is still voluntary, by 2010 all EU states are supposed to have implemented the EPBD.*

One of the key driving forces of European energy-efficient design is the EU's 2002 Directive.† Each of the 27 member states of the European Union is responsible for individual implementation of the EPBD through national laws.‡ The main focus of European sustainable building design at this time is on reducing energy use directly and carbon emissions indirectly.

The EPBD has five main themes:

- *Certificates.* When buildings are constructed, sold, or rented out, the owner must provide an energy performance certificate to the prospective buyer or tenant.
- *Inspection.* All large boilers and air-conditioning units must be inspected to ensure proper operation.
- *Experts.* Qualified experts must carry out inspections and provide the analysis for the certificates.

*See the more comprehensive discussion in Jerry Yudelson, *Green Building Trends: Europe*, 2009, Washington, DC: Island Press, Chapters 3 and 9.

†http://www.buildup.eu/home/professional, accessed July 20, 2009.

‡http://europa.eu/abc/european_countries/others/index_en.htm, accessed August 28, 2008.

- *Calculations.* Energy use estimates must be made with adequate software whose calculation methods are transparent and widely accepted.
- *Minimum energy performance requirements.* For new buildings with a total useful floor area greater than 1000 square meters (10,760 square feet), each country shall require that every project consider the technical, environmental, and economic feasibility of alternative systems such as decentralized energy supply systems based on renewable energy, combined heat and power systems, district heating or cooling (if available), and heat pumps (under certain conditions). In addition, owners of existing buildings greater than 1000 square meters that undertake major renovations shall be required to upgrade their energy performance to meet new minimum requirements.

This is a comprehensive program that we should emulate in the United States and Canada. Without clear standards for new buildings and major upgrade requirements for existing buildings, we'll never be able to significantly reduce the carbon emissions from residential and commercial buildings.

Driving the EPBD are the heightened European concern over the role of building energy and materials use in global carbon dioxide production, constraints on energy supplies, and the potential for catastrophic changes in the global climate as a result of increased carbon dioxide concentrations in the earth's atmosphere. To date, European national governments have been far more willing to accept the conclusions of climate science than American or Canadian governments and have been willing to take that science and develop practical public policies for reversing the growth of carbon emissions, including subsidies, laws, and regulations to implement these policies.

As for future development of the EPBD in Europe, Bill Bordass of the U.K.'s Usable Buildings Trust provides his assessment of the situation:*

> The directive is currently under review in Brussels. Written submissions were invited earlier this year [2008] and are being collated in preparation for a process that will lead to revisions in 2010. To avoid any confusion, the European Commission plans not just to make amendments, but to publish a new version of the directive. I suspect that this will require Display Energy Certificates (DECs) on a wider range of commercial buildings and possibly all.† DECs may also need to be based on actual energy use. The new requirements would probably come fully into force two to three years after 2010.

THE UNITED KINGDOM EXAMPLE

The United Kingdom has been a leader in implementing the EPBD. The government introduced requirements for Energy Performance Certificates (EPCs) for new buildings

---

*Interview with Bill Bordass, July 2008.

†In Europe there are two forms of energy certificate. Energy performance certificates (EPCs) are designed to inform property transactions; they use "asset ratings" (based on theoretical calculations) and provide a standardized basis for comparing buildings on the market, be they old, new, or refurbished. Display energy certificates (DECs) are for display in public buildings. Countries can choose whether to base their DECs on calculated asset ratings (as in Scotland) or on measured annual energy use in the form of operational ratings (as in England, Wales, and Northern Ireland).

and DECs for existing buildings. Each building is graded from A (best) to G (worst), with a "G" representing something you wouldn't even want to tell your spouse or kids about, a building that might use 50 percent more energy than the average of its type. The current average building energy use lies between D and E.

The expected timetable of regulation in the United Kingdom follows. Most of these dates deal with the EPCs. Initially, the primary difficulty in the United Kingdom appeared to be training enough qualified assessors to make the required analyses:* (Obviously, this is a problem that solves itself over time.)

- *April 2008*. EPCs required on construction for all new dwellings. EPCs required for the construction, sale, or rent of buildings, other than dwellings, with a floor area greater than 10,000 square meters (107,000 square feet)
- *July 2008*. EPCs required for the construction, sale, or rent of buildings, other than dwellings, with a floor area greater than 2500 square meters (27,000 square feet)
- *October 2008*. EPCs required on the sale or rent of all remaining dwellings. EPCs required on the construction, sale, or rent of all remaining buildings, other than dwellings. Display energy certificates required for all public buildings larger than 1000 square meters (10,700 square feet)
- *January 4, 2009*. First inspection of all existing air-conditioning systems larger than 250 kilowatts must be completed
- *January 4, 2011*. First inspection of all remaining air-conditioning systems larger than 12 kilowatts must be completed

What does a building owner get from an EPC, except compliance with the law and an expensive invoice from the consultant performing the analysis?

An EPC will provide an energy rating for a building that is based on the performance potential of the building itself (the fabric) and its services (such as heating, ventilation, and lighting). The certificate provides an energy rating of the building from A to G, where A is very efficient and G is the least efficient. The energy performance of the building is shown as a carbon-dioxide-based index. The EPC is accompanied by a report that provides recommendations on using the building more effectively, cost-effective improvements to the building, and other more expensive improvements which could enhance the building's energy performance.†

The DECs will work much the same way, except that they will use the actual energy performance data of existing buildings. The difficulty is to establish an effective way to compare buildings fairly and generate a grade based not just on square footage but also on the number of occupants, daily and yearly occupancy schedule, presence or absence of a data center, and similar considerations.

---

*Building (U.K.) magazine, www.building.co.uk/sustain_story.asp?storycode=3116537&origin=bldgsustainnewsletter, accessed August 27, 2008.

†Ibid.

III

INTERVIEWEES

Michael Arny, President, Leonardo Academy
Stephen Ashkin, President, The Ashkin Group
Elaine Aye, Principal, Green Building Services
Brodie Bain, Principal, Mithun
Mark Bettin, Vice President—Engineering, Merchandise Mart Properties
Jack Beuttell, Sustainability Manager, Hines
Denis Blackburne, Chief Financial Officer, Melaver
Ron Blagus, Energy Market Director, Honeywell Building Solutions
Bill Bordass, Principal, William Bordass Associates
Clark Brockman, Associate Principal and Director of Sustainability Resources, SERA Architects
Barbara Cielsa, Leader of the Sustainable Design Consulting Group, HOK
Margot Crossman, General Manager, Unico Properties
Peter Dahl, Sustainability Specialist, Sebesta Blomberg
Mark Frankel, Technical Director, New Buildings Institute
Lisa Galley, Managing Principal, Galley Eco Capital
Barry Giles, CEO, Building Wise
Paul Goldsmith, Operations/Sustainability Champion, Harley Ellis Devereaux
Darren Goody, Commissioning and Auditing Manager, Green Building Services
Gretchen Hancock, Project Manager for Corporate Environmental Programs, General Electric
Helee Hillman, Senior Project Manager, Jones Lang LaSalle
Don Horn, Director, Sustainability Program in the Office of Federal High-Performance Green Buildings, U.S. General Services Administration
Brad Jones, Division Leader for Commissioning and Sustainability, Sebesta Blomberg
Wade Lange, Vice President of Property Management, Ashforth Pacific
Jerry Lea, Senior Vice President, Hines

Tony Litton, President, Sebesta Blomberg

Mychele Lord, President, LORD Environmental Strategies

Mike Lyner, Principal, RSP i-space

Andrew McAllen, Senior Vice President, Real Estate Management, Oxford Properties Group

Paul McCown, Senior Project Manager, SSRCx

Wayne Mezick, Chief Engineer, Ashforth Pacific

Mark Morello, President, Infinity Water Management

Stefan Mühle, General Manager, Orchard Hotel & Orchard Garden Hotel

Scott Muldavin, Executive Director, Green Building Finance Consortium

Doug Norwood, Senior Mechanical Engineer, Sacramento Municipal Utility District

Derick Podratz, Manager and Group Leader, Sebesta Blomberg

Rick Pospisil, Director of Facilities, USAA Real Estate Company

Wesley Powell, Managing Director, Jones Lang LaSalle

Jim Rock, Senior Vice President, Leasing, Unico Properties

Paul Rode, Project Manager, Johnson Controls

Adam Rose, Property Manager, Hines

Leo Roy, Director of Environmental and Energy Services, Vanasse Hangen Brustlin

Thomas Saunders, Special Projects Manager, BRE Global

Dana Schneider, Northeast Market Lead for Energy and Sustainability, Jones Lang LaSalle

Craig Sheehy, President, Envision Realty Services

C. Johnathan Sitzlar, Supervisory Building Manager, U.S. General Services Administration

Al Skodowski, Senior Vice President, Director of LEED and Sustainability, Transwestern

Joel Stout, LEED Sustainability Specialist, Sebesta Blomberg

Gary Thomas, Director, Sustainability Program, CB Richard Ellis

Theresa Townsend, Senior Architect, Division of the State Architect, California Department of General Services

Stephen Zanolini, Global Facility Manager, Caterpillar Financial

IV

RATING SYSTEMS FOR GREENING EXISTING BUILDINGS

This appendix presents detailed profiles of the three most developed rating systems for greening existing buildings: Australia, the United Kingdom, and the United States. I've also included a brief mention of LEED Canada for Existing Buildings, since that system is likely to be adopted and in place by the end of 2009. I expect most other rating systems for existing buildings that develop around the world will assess similar environmental attributes for greening existing buildings, though the weightings of various actions may differ from country to country.

Australia

The Green Building Council of Australia (GBCA) began developing Green Star as a rating system for evaluating the environmental design of buildings in 2002. The GBCA promotes green building programs, technologies, design practices, and operations. Rating tools are currently available or in development for most building market segments, including commercial offices, retail, schools, universities, multiunit residential buildings, industrial facilities, and municipal buildings.*

The GBCA released the pilot version of their Green Star: Office Existing Building rating system in April 2007. The goal of this rating system is to assess the current environmental potential of existing buildings. Property managers will find it useful when identifying upgrade and retrofit priorities. The rating system also assists corporate sustainability and environmental reporting efforts. For a project to be eligible for existing building certification under Green Star, at least 24 months must have passed since the issuance of the construction certificate.†

*http://www.gbca.org.au/about/, accessed April 15, 2009.

†http://www.gbca.org.au/uploads/254/1534/Fact%20Sheet%20Green%20Star%20-%20Office%20Existing%20Building%20041208.pdf, accessed April 15, 2009.

TABLE IV.1 GREEN STAR—OFFICE EXISTING BUILDING (EXTENDED PILOT) CATEGORIES AND WEIGHTINGS

ENVIRONMENTAL IMPACT CATEGORY	WEIGHTS
Management	20%
Indoor Environment Quality	20%
Energy	25%
Transport	10%
Water	12%
Materials	4%
Land Use & Ecology	4%
Emissions	5%
Total	100%

Every Green Star rating tool is organized into eight environmental impact categories and an innovation category. Credits are awarded within each of the categories, depending on a building's environmental performance and characteristics. Points are achieved when specified actions for each credit are successfully performed and/or demonstrated. Table IV.1 outlines the categories and weightings within the existing building rating system.*

The number of credits for each category is totaled and a percentage score is calculated as follows:†

$$\text{Category score (\%)} = (\text{total number of points achieved/total number of points available}) \times 100$$

Environmental weighting is applied to each category score, which balances the inherent weighting that occurs through the differing number of points available in each category. The weights reflect issues of environmental importance for each state or territory of Australia, and thus differ by region. The weighted category score is calculated as follows:‡

$$\text{Weighted category score (\%)} = \text{category score (\%)} \times \text{weighting factor (\%)}/100$$

The sum of the weighted category scores, plus any innovation points, determines a project's rating. Only buildings that achieve a rating of four stars and above are

*Ibid.

†http://www.gbca.org.au/green-star/green-star/green-star-rating-calculation/1542.htm, accessed November 5, 2008.

‡Ibid.

TABLE IV.2 GREEN STAR—OFFICE EXISTING BUILDING (EXTENDED PILOT) CERTIFIED RATINGS

SCORE	RATING	STAR RATING
10–19	Acceptable	*
20–29	Average Practice	**
30–44	Good Practice	***
45–59	Best Practice	****
60–74	Australian Excellence	*****
75–100	World Leadership	******

certified by the GBCA. The rating levels and their respective scores are listed in Table IV.2.*

Canada

The Canada Green Building Council seeks to transform the built environment by developing best design practices and guidelines for green building. The Council has adapted the U.S. Green Building Council's (USGBC) Leadership in Energy and Environmental Design (LEED) rating system to Canadian climates, construction practices, and regulations (see the section, "United States"). Currently LEED® Canada has developed rating systems to certify New Construction, Commercial Interiors, and Core and Shell projects.†

LEED Canada for Existing Buildings: Operations and Maintenance rating system is scheduled to begin accepting project registrations in August 2009.‡

As with LEED in the United States, the prerequisites and credits in the LEED Canada system are organized into five principal categories:§

- Sustainable Sites
- Water Efficiency
- Energy and Atmosphere
- Materials and Resources
- Indoor Environmental Quality
- Innovation and Design Process

*http://www.gbca.org.au/uploads/254/1534/Fact%20Sheet%20Green%20Star%20-%20Office%20Existing%20Building%20041208.pdf, op.cit.

†http://www.cagbc.org/database/rte/LEED_Certified_Projects_in_Canada_Updated_090319.pdf, accessed April 23, 2009.

‡http://www.cagbc.org/leed/systems/existing_buildings/index.php, accessed July 20, 2009.

§http://www.cagbc.org/leed/what/index.php, accessed November 7, 2008.

Project ratings are determined by the number of points awarded for the successful completion of credit requirements. Depending on the number of points awarded, there are four possible levels of certification:*

- Certified
- Silver
- Gold
- Platinum

United Kingdom

The Building Research Establishment (BRE) is the leading authority on sustainable design in the United Kingdom. In the 1990s, BRE developed the BRE Environmental Assessment Method (BREEAM), which assesses buildings against various sustainability criteria and provides an overall score that falls within a certain rating level. BREEAM can be used as an environmental assessment tool for any type of building, in the United Kingdom or internationally. This system can be applied to single buildings or entire portfolios, and it can also be tailored for various stages in the life cycle of a building. For instance, BREEAM In-Use (BIU) is designed to help building managers reduce operating costs and improve the environmental performance of existing buildings.[†]

BIU is compatible with environmental standards and regulations in the United Kingdom, such as Energy Performance Certificates (EPCs), Display Energy Certificates (DECs), and the Regulatory Reform (Fire Safety) Order. It is also compatible with environmental performance and reporting systems like the Investment Property Databank Environment Code, ISO 14001, and the Global Reporting Initiative (GRI).[‡]

The BIU standard breaks the environmental performance of a building into three parts:

- Asset (Part 1)—the performance characteristics of the building itself
- Building Management (Part 2)—the performance of policies, procedures, and practices related to the operation of the building
- Organizational Effectiveness (Part 3)—the understanding and implementation of building operations and management policies

Part 1 can be assessed independently of Parts 2 and 3. There is an online tool that guides property managers and owners through a self-assessment of the BIU standard. BREEAM uses licensed, third-party assessors to evaluate BIU assessment reports.[§]

---

*Ibid.

[†]http://www.breeam.org/filelibrary/SD096__Rev_0__BREEAM_In_Use_Scheme_Document.pdf, accessed April 14, 2009.

[‡]http://www.breeam.org/filelibrary/BREEAM_In-Use_Standard_BES_5058_Issue_1.pdf, accessed April 14, 2009.

[§]Ibid.

TABLE IV.3 BIU SECTIONS AND WEIGHTS*

SECTIONS	PART 1 SECTION WEIGHTING	PART 2 SECTION WEIGHTING	PART 3 SECTION WEIGHTING
Energy	26.5%	31.5%	19.5%
Water	8.0%	5.5%	3.5%
Materials & Waste	13.5%	7.5%	16.0%
Health & Wellbeing	17.0%	15.0%	15.0%
Pollution	14.0%	13.0%	10.5%
Transport	11.5%	0.0%	18.5%
Land Use & Ecology	9.5%	12.5%	5.0%
Management	0.0%	15.0%	12.0%
Totals	100%	100%	100%

*Ibid.

There are numerous assessment criteria for which a project can receive credits. The criteria are organized into sections and are weighted differently for each part of the standard. Table IV.3 lays out the criteria sections and weights for each part.

The number of credits awarded for each section are summed and compared to total number of credits available as follows:*

$$\text{Section score (\%)} = (\text{total number of points achieved/total number of points available}) \times 100$$

The resulting percentage is then weighted (according to the weights listed in Table IV.1) as follows:

$$\text{Weighted section score (\%)} = \text{section score (\%)} \times \text{weighting factor/100}$$

This produces a weighted section score. These section scores are summed, as well as any innovation credits that have been achieved. This produces the final BIU score, which translates to a particular rating level or number of stars. A licensed assessor completes a review and verification of the assessment report and generates a BIU rating certificate based on the assessment. The rating levels and their respective scores are listed in Table IV.4.†

*http://www.breeam.org/filelibrary/SD096__Rev_0__BREEAM_In_Use_Scheme_Document.pdf, accessed April 14, 2009.

†http://www.breeam.org/filelibrary/BREEAM_In-Use_Standard_BES_5058_Issue_1.pdf, op.cit.

TABLE IV.4 BIU SCORING CATEGORIES AND RATING LEVELS

ASSESSMENT SCORE (%)	RATING	STAR RATING
<10	Unclassified	-
10–25	Acceptable	*
25–40	Pass	**
40–55	Good	***
55–70	Very Good	****
70–85	Excellent	*****
>85	Outstanding	******

United States

The U.S. Green Building Council (USGBC) introduced the LEED Green Building Rating System in version 2.0 in 2000. Since its inception the system has evolved and expanded and is now considered a leading method of measuring and rating building performance in many countries of the world. There are nine different rating systems that apply to particular building market segments or project types. This appendix focuses on the following LEED rating systems:

- LEED for New Construction (LEED-NC)
- LEED for Commercial Interiors (LEED-CI)
- LEED for Existing Buildings: Operations & Maintenance (LEED-EBOM)

All LEED rating systems were recently updated through LEED v3, a multifaceted initiative that involved streamlining and increasing capacity for project execution, documentation, and certification. Several key advancements were made through this initiative: the rating systems were updated and revised, the prerequisites and credits were better aligned and reweighted, and regional environmental priorities were added. One of the biggest changes made through LEED v3 is the incorporation of LEED 2009, which changes the weighting assigned to various LEED credits. The available points in the earlier LEED rating systems were redistributed so that a given credit's point value "more accurately reflects its potential to either mitigate the negative or promote the positive environmental aspects of a building."* In calculating the weightings between various credit points, the USGBC used an environmental weighting method developed by the U.S. Environmental Protection Agency.

*http://www.usgbc.org/ShowFile.aspx?DocumentID=4121, accessed November 4, 2008.

LEED-NC is intended to guide the design and construction/renovation of high-performance commercial and institutional projects, along with residential projects four stories and higher. These projects include, but are not limited to, government offices, retail and service establishments, and commercial offices. This rating system provides a set of performance standards that ensures certified buildings are healthy, durable, and environmentally sound.

LEED-CI is a set of performance standards intended to guide tenant improvements. This rating system is typically used in office, retail, restaurant, health care, hotel, and educational settings. In the LEED system, tenants are defined as lease holders or occupants who pay rent to use a building. LEED-CI is typically used for remodels or situations in which neither the building envelope nor the HVAC system would be substantially altered.

LEED-EBOM helps building owners and operators measure impacts of operations, improvements, and maintenance on a consistent scale. The goal for project teams employing this rating system is to maximize operational efficiency while minimizing environmental impacts. This rating system also allows for ongoing certification for buildings throughout their lifetime. Buildings can be recertified every one to five years under this system.

Table IV.5 compares the LEED 2009 versions of the above rating systems. Certification levels and total possible points for each credit category are compared at the bottom of the table. [Asterisks (*) indicate cumulative points. Innovation points can be earned for exceptional performance in most of the credit categories with cumulative points.]

TABLE IV.5 COMPARISON OF THE THREE MAIN LEED RATING SYSTEMS, 2009 VERSION

LEED CREDIT CATEGORY	LEED-NC POINTS	LEED-CI POINTS	LEED-EBOM POINTS
Sustainable Sites			
Prerequisite Construction Activity Pollution Prevention	Required	–	–
LEED Certified Design and Construction	–	Up to 5	4
Building Exterior and Hardscape Management Plan	–	–	1
Integrated Pest Management, Erosion Control, and Landscape Management Plan	–	–	1
Site Selection	1	–	–
Development Density & Community Connectivity	5	6	–
Brownfield Redevelopment	1	–	–

(*Continued*)

LEED CREDIT CATEGORY	LEED-NC POINTS	LEED-CI POINTS	LEED-EBOM POINTS
Alternative Commuting Transportation	–	–	3 to 15*
10% Reduction			3*
25% Reduction			4*
50% Reduction			4*
75% Reduction or greater			4*
Alternative Transportation	Up to 12	Up to 10	–
Public transportation access	6	6	
Bicycle storage and commuting	1	2	
Low-emitting and fuel-efficient vehicles	3	–	
Parking capacity	2	2	
Delivery service	–	–	
Incentives	–	–	
Car-share membership	–	–	
Alternative transportation education	–	–	
Site Development		–	
Protect or restore habitat	1		1
Maximize open space	1		–
Stormwater Design		–	
Quantity control	1		1
Quality control	1		–
Heat Island Effect		–	
Nonroof	1		1
Nonroof 25% shade	–		–
Nonroof 50% shade	–		–
Nonroof 75% shade	–		–
Roof	1		1
Light Pollution Reduction	1	–	1
Tenant Design & Construction Guidelines	–	–	–
Water Efficiency			
Prerequisite Minimum Indoor Plumbing Fixture and Fitting Efficiency	–	–	Required
Water Performance Measurement	–	–	
Whole building metering			1*
Submetering			1*

LEED CREDIT CATEGORY	LEED-NC POINTS	LEED-CI POINTS	LEED-EBOM POINTS
Additional Indoor Plumbing Fixture and Fitting Efficiency	–	–	Up to 5
10% Reduction			1*
15% Reduction			1*
20% Reduction			1*
25% Reduction			1*
30% Reduction			1*
Water Efficient Landscaping	Up to 4	–	Up to 5
50% Reduction	2*		1*
62.5% Reduction	–		1*
75% Reduction	–		1*
87.5% Reduction	–		1*
100% Reduction	2*		1*
Innovative Wastewater Technologies	2	–	–
Water Use Reduction	Up to 4	Up to 11	–
20% Reduction	Required	Required	
30% Reduction	2*	6*	
35% Reduction	1*	2*	
40% Reduction	1*	3*	
Cooling Tower Water Management	–	–	
Chemical management			1
Non-potable water source use			1
Energy & Atmosphere			
Prerequisite Energy Efficiency Best Management Practices	–	–	Required
Prerequisite Fundamental Commissioning of the Building Energy Systems	Required	Required	–
Prerequisite Minimum Energy Performance	Required	Required	Required (ENERGY STAR rating 69)
Prerequisite Fundamental Refrigerant Management	Required	Required	Required
Optimize Energy Performance	Up to 19	Up to 5	Up to 18
Optimize Energy Performance, Lighting Controls	–	Up to 3	–

(*Continued*)

LEED CREDIT CATEGORY	LEED-NC POINTS	LEED-CI POINTS	LEED-EBOM POINTS
Optimize Energy Performance, HVAC	–	5 to 10	–
Optimize Energy Performance, Equipment, and Appliances	–	Up to 4	–
Existing Building Commissioning	–	–	
Investigation and analysis			2
Implementation			2
Ongoing commissioning			2
Renewable Energy	Up to 7	–	Up to 6
Enhanced Commissioning	2	5	–
Enhanced Refrigerant Management	2	–	1
Performance Measurement & Verification	3	2 to 5	
Building automation system			1
System level metering			
40%			1*
80%			1*
Base building			
Tenant submetering			
Green Power	2	5	–
Emissions Reduction Reporting	–	–	1
Materials & Resources			
Prerequisite Sustainable Purchasing Policy	–	–	Required
Prerequisite Solid Waste Management Policy	–	–	Required
Prerequisite Storage & Collection of Recyclables	Required	Required	–
Sustainable Purchasing	–	–	
Ongoing consumables 40% of purchases			1
Durable goods, electric			1
Durable goods, furniture			1
Facility alterations and additions			1
Reused mercury in lamps 90 pg/lum-h			1
Food			1
Building Reuse	Up to 4		–
Tenant space, long-term commitment	–	1	
Maintain 40% of interior nonstructural components	–	1*	

LEED CREDIT CATEGORY	LEED-NC POINTS	LEED-CI POINTS	LEED-EBOM POINTS
Maintain 60% of interior nonstructural components	–	1*	
Maintain 25% of existing walls, floors, and roof			
Maintain 33% of existing walls, floors, and roof			
Maintain 42% of existing walls, floors, and roof			
Maintain 50% of existing walls, floors, and roof			
Maintain 55% of existing walls, floors, and roof	1*		
Maintain 75% of existing walls, floors, and roof	1*		
Maintain 95% of existing walls, floors, and roof	1*		
Maintain 50% of interior nonstructural elements	1*		
Construction Waste Management	Up to 2		–
Divert 50% from disposal	1*	1*	
Divert 75% from disposal	1*	1*	
Solid Waste Management	–	–	
Waste stream audit			1
Ongoing consumables 50% waste diversion			1
Durable goods			1
Facility alterations and additions			1
Materials Reuse	Up to 2		–
5% salvaged, refurbished, or reused materials	1*	1*	
10% salvaged, refurbished, or reused materials	1*	1*	
Reuse 30% of furniture and furnishings	–	1	
Recycled Content	Up to 2		–
10% (postconsumer + 1/2 preconsumer)	1*	1*	
20% (postconsumer + 1/2 preconsumer)	1*	1*	
Regional Materials	Up to 2	Up to 2	–
10% extracted, processed, and manufactured regionally	1*	1*	
20% extracted, processed, and manufactured regionally	1*	1*	
Rapidly Renewable Materials	1	1	–
Certified Wood	1	1	–

(*Continued*)

LEED CREDIT CATEGORY	LEED-NC POINTS	LEED-CI POINTS	LEED-EBOM POINTS
Indoor Environmental Quality			
Prerequisite Minimum IAQ Performance	Required	Required	Required
Prerequisite Environmental Tobacco Smoke (ETS) Control	Required	Required	Required
Prerequisite Green Cleaning Policy	–	–	Required
IAQ Management Program	–	–	1
Outdoor Air Delivery Monitoring	1	1	1
Increased Ventilation	1	1	1
Reduce Particulates in Air Distribution	–	–	1
Construction IAQ Management Plan			
During construction	1	1	1
Before occupancy	1	1	–
Occupant Comfort	–	–	1
Low-Emitting Materials	Up to 4	Up to 5	–
Adhesives and sealants	1	1	
Paints and coatings	1	1	
Flooring	1	1	
Composite wood and agrifiber products	1	1	
Furniture	–	1	
Ceiling and wall systems	–	–	
Indoor Chemical & Pollutant Source Control	1	1	–
Controllability of Systems, Lighting	1	1	1
Controllability of Systems, Thermal Comfort	1	1	–
Thermal Comfort			1
Design	1	1	–
Employee verification	1	1	–
Compliance	–	–	–
Daylight & Views			
50% Daylight/45% views	–	–	1
Daylight for 75% of spaces	1*	1*	
Daylight for 90% of spaces	–	1*	
Views for 90% of spaces	1	1	

LEED CREDIT CATEGORY	LEED-NC POINTS	LEED-CI POINTS	LEED-EBOM POINTS
Green Cleaning	–	–	
High performance cleaning program			1
Custodial effectiveness assessment			1
Sustainable cleaning products and materials purchases			1
Sustainable cleaning equipment			1
Indoor chemical and pollutant source control			1
Indoor integrated pest management			1
Innovation & Design Process			
Innovation in Design	Up to 5	Up to 5	Up to 4
LEED Accredited Professional	1	1	1
Documenting Sustainable Building Cost Impacts	–	–	1
Regional Priority Credits			
Regional Priority Credits	Up to 4	Up to 4	Up to 4
Point Totals			
Total possible points for Sustainable Sites	26	21	26
Total possible points for Water Efficiency	10	11	14
Total possible points for Energy & Atmosphere	35	37	35
Total possible points for Materials & Resources	14	14	10
Total possible points for Indoor Environmental Quality	15	17	15
Total possible points for Innovation & Design Process	6	6	6
Total possible points for Regional Priorities	4	4	4
Total rating system points available	110	110	110
Certification levels (minimum points)			
Certified	40	40	40
Silver	50	50	50
Gold	60	60	60
Platinum	80	80	80

*Cells indicate credits where points are cumulative for increasing levels of achievement.

V

RESOURCES FOR FURTHER INFORMATION

Books

Lippe, P., ed. (2008), *Lessons Learned: The Costs and Benefits of High-Performance Buildings.* Vol. 5, Earth Day New York, New York.

- A valuable compilation of articles on green building by leading authorities.

Melaver, M. and Mueller, P. (2008), *The Green Building Bottom Line*. McGraw-Hill Professional, New York.

- Provides insights and data that demonstrate the true costs and benefits of building green, including conversions of existing buildings.

U.S. Green Building Council (2009), *LEED Reference Guide for Green Building Operations and Maintenance 2009 Edition*. U.S. Green Building Council.

- The essential reference for anyone concerned with greening existing buildings.

Yudelson, J. (2009), *Green Building Trends: Europe.* Island Press, Washington, DC.

- Provides a current picture of the European approach to greening existing and new buildings, with a review of the green building rating systems in many countries.

Yudelson, J. (2009), *Sustainable Retail Development: New Success Strategies*. Springer, Dordrecht, the Netherlands.

- The most up-to-date comprehensive survey of sustainable initiatives in the retail sector, worldwide, with a focus on North America and Western Europe. Shows how retailers and developers are greening existing operations.

Yudelson, J. (2008), *Green Building through Integrated Design.* McGraw-Hill Professional, New York.

- Provides many insights into how to organize building teams to deliver high-performance results. Though aimed at new construction, many of the lessons in team building and project management are directly applicable to existing building upgrades.

Yudelson, J. (2007), *The Green Building Revolution.* Island Press, Washington, DC.

- The first comprehensive overview and manifesto of the green building movement. Covers the costs of green building, the business case, and a review of each major market sector.

Research Articles

Eichholtz, P.; Kok, N.; and Quigley, J., 2009, "Doing Well By Doing Good? An Analysis of The Financial Performance of Green Office Buildings in The USA," *RICS Research Report*, March 2009, London.

- A solid academic review of the business case benefits of about 900 green buildings in the United States.

Lawrence Berkeley National Laboratory, 2004, "The Cost-Effectiveness of Commercial Buildings Commissioning." Available at: http://eetd.lbl.gov/emills/PUBS/Cx-Costs-Benefits.html

- The most comprehensive review of the benefits of building commissioning, a "meta-study" of more than 120 separate commissioning studies.

Leonardo Academy, 2008, "The Economics of LEED for Existing Buildings, 2008 Edition," April 21, 2009. Available at: http://www.leonardoacademy.org/download/Economics%20of%20LEED-EB%2020090222.pdf

- A detailed survey of the benefits of LEED-EB, by a leading nonprofit in the field.

Loftness, V.; Hartkopf, V.; Gurtekin, B.; Hua, Y.; Snyder, M.; Gu, Y.; and Xiaodi Yang Graduate Students, Building Investment Decision Support (BIDS™) (Pittsburgh: Center for Building Performance and Diagnostics, Carnegie Mellon University, n.d.). Available at: http://cbpd.arc.cmu.edu/ebids, accessed August 21, 2009.

- The premier collection of information on academic studies of the benefits of daylighting, indoor air quality, and other green building measures.

Miller, N.; Spivey, J.; and Florance, A., 2008, "Does Green Pay Off?" July 8, 2008, CoStar. Available at: http://www.costar.com/josre/pdfs/CoStar-JOSRE-Green-Study.pdf

- The first comprehensive analysis of the benefits of LEED and ENERGY STAR buildings, using a database of more than 1300 buildings.

Pacific Northwest National Laboratory, 2008, "Assessing Green Building Performance: A Post-Occupancy Evaluation of 12 GSA Buildings." Available at: www.gsa.gov/gsa/cm_attachments/GSA_DOCUMENT/GSA_WBDG_Report_Final_R2-p-q5Q_0Z5RDZ-i34K-pR.pdf

- This report analyzes how sustainably designed buildings perform in comparison to traditionally designed buildings. Energy, water, maintenance and operations, waste, recycling, transportation, and occupancy satisfaction metrics are included.

Pivo, G. and Fisher, J., 2009, "Investment Returns From Responsible Property Investments, Energy-Efficient, Transit-oriented and Urban-Regeneration Office Properties in the U.S. from 1998–2008," Working Paper, Responsible Property Investing Center, Boston College and University of Arizona, Benecki Center for Real Estate Studies, Indiana University; May 14, 2009. Available at: www.u.arizona.edu/~gpivo/

- Shows how green buildings and responsible property investing lead to better business outcomes for property owners.

U.S. General Services Administration, March 2009, "Energy Savings and Performance Gains in GSA Buildings: Seven Cost-effective Strategies." Available at: www.gsa.gov/graphics/pbs/GSA_SevenStrategies_090327screen.pdf, accessed August 21, 2009.

- The GSA's Workplace Performance Study identifies seven key areas that offer the potential for significant energy reduction and savings. The study encompasses survey results from over 6000 federal workers and environmental conditions at 624 workstations in 22 separate buildings.

Urban Land Institute, 2008, "Retrofitting Buildings to be Green," Washington, DC.

- A soft-bound compilation of articles and reports, project descriptions, and sections from other ULI books on the subject of greening existing buildings.

Periodicals

Architectural Products (www.arch-products.com)

- Provides monthly coverage of green products and product application information for architects, designers, and product specifiers involved in commercial and institutional building design.

Architectural Record (http://archrecord.construction.com)

- An excellent source of green building information for the mainstream architectural community and a good way for engineers to keep up with the evolving discussion of sustainability among architects.

Building Design & Construction (www.bdcnetwork.com)

- BD&C is one of the authoritative voices in the industry. Written primarily for "Building Team" practitioners, it is eminently accessible to anyone. More editorial focus each year on greening existing buildings.

Building Operating Management (www.facilitiesnet.com)

- BOM is an essential resource for building operators and facility managers. Covers practical topics in building management and operations, with a strong focus on energy management.

Consulting-Specifying Engineer (www.csemag.com)

- A monthly trade magazine for engineering management professionals. Strong focus on practical energy management and energy retrofits.

Eco-Structure Magazine (www.eco-structure.com)

- Eco-structure is the best illustrated of the trade magazines covering the green building industry. Good case studies and a broad selection of topics make it a good read for keeping up.

Environmental Design & Construction (www.edcmag.com)

- Now 10 years old, ED&C provides first-class editorial coverage of the relevant issues for greening existing buildings, along with well-written case studies of leading green building projects.

Green Lede (www.costar.com/news)

- Green Lede is a CoStar feature column examining green building and its impact on commercial real estate, politics, and the environment. A great way to keep in touch with the world of commercial real estate green thinking.

GreenSource Magazine (http://greensource.construction.com)

- From the publishers of Engineering News-Record and Architectural Record, the most authoritative publications in their respective fields, this magazine covers the field of green buildings. The case studies are the best you'll find written anywhere.

High-Performance Buildings (www.hpbmagazine.org)

- This quarterly magazine is published by ASHRAE and distributed to building owners, facility managers, architects, contractors, and engineers. The goal is to help decision makers in the building community learn about the benefits of innovative technologies and energy-efficient design and operation. There are excellent case studies of exemplary buildings, with strong technical details.

Journal of Green Building (www.collegepublishing.us/journal.htm)

- The best peer-reviewed quarterly serving the green building industry, with both industry-relevant and academic research.

Solar Today (www.solartoday.org)

- The official publication of the American Solar Energy Society, but written for a general audience. Strong technical content with a lot of coverage of the commercial sector and solar retrofits.

Web Sites and Organizations

American Council for an Energy-Efficient Economy (www.acee.org)

- A nonprofit organization that conducts in-depth technical and policy analysis on energy efficiency and economic issues.

American Institute of Architects Committee on the Environment (http://www.aia.org/practicing/groups/kc/AIAS074686)

- This committee's annual Top Ten Green Projects is the profession's best-known recognition program for sustainable design excellence. Most of the focus is on new buildings.

American Society of Heating, Refrigerating and Air-Conditioning Engineers (ASHRAE, www.ashrae.org)

- International technical organization that advances heating, ventilation, air conditioning, and refrigeration to serve humanity and promote a sustainable world. Publishes *High Performance Buildings Magazine*.

Association for the Advancement of Sustainability in Higher Education (AASHE) (www.aashe.org)

- Provides a comprehensive overview of sustainable programs and activities in American higher education, including LEED certifications and climate-neutral commitments and programs.

BetterBricks (www.betterbricks.com)

- An excellent resource for energy-efficient and green building design from the Northwest Energy Efficiency Alliance (www.nwalliance.org), a utility-funded organization that offers hundreds of articles, interviews, and technical resources for sustainable design.

Building Energy Performance News (www.bepinfo.com)

- A good daily compendium of every going on in the world of building energy performance.

Building Owners and Managers Association's Energy Efficiency Program (BEEP©), http://www.boma.org/TrainingAndEducation/BEEP/Pages/default.aspx)

- A targeted educational program for commercial real estate professionals on proven no- and low-cost strategies for optimizing building energy performance. The program was developed in partnership with the EPA's ENERGY STAR program.

Canada Green Building Council (www.cagbc.org)

- Canada's own Green Building Council (see below), closely related to the U.S. Green Building Council. Resources and rating system guides can be downloaded from their site.

Commercial Building Tax Deduction Coalition (www.efficientbuildings.org)

- This site covers the Energy Policy Act of 2005 commercial energy efficiency tax deduction.

Database of State Incentives for Renewables & Efficiency (www.dsireusa.org)

- Provides nationwide, interactive maps for incentives ad rebates at the state and federal levels.

Efficiency Vermont (www.efficiencyvermont.com)

- Provides technical assistance and information on financial incentives to help households and businesses in Vermont reduce their energy costs.

Environmental Building News (www.buildinggreen.com)

- Offers print and electronic resources designed to help building industry professionals and policy makers improve the environmental performance, and reduce the adverse impacts, of buildings. This is a long-standing and outstanding resource for understanding the full range of green building options.

Environmental Leader (www.environmentalleader.com)

- A daily newsletter, archived on the Web site, tailored to corporate executives that want to stay on top of energy, environmental, and sustainability news. Good coverage of corporate initiatives that involve greening existing buildings.

Green Building Certification Institute (www.gbci.org)

- GBCI is the official organization that credentials professionals and oversees the LEED certification process. Created in 2008 as a spinoff from the USGBC. If you want to stay on top of what's happening with LEED-EB certification, you'll have to check this Web site frequently.

Green Building Finance Consortium (www.greenbuildingfc.com)

- The Green Building Finance Consortium (GBFC) is a group of leading corporations, real estate companies, and trade groups who have joined together to address the need for independent research and analysis of investment in green or energy-efficient buildings.

GreenerBuildings.com

- A good Web site for a continuous read of the sustainable building movement. Just about every major initiative is covered here, as soon as it happens.

Green Guide for Health Care (www.gghc.org)

- Publishes a "best practices" guide for health care facility design, construction, and operations. Lots of practical information relevant to the health care sector.

New York State Energy Research and Development Authority (NYSERDA, www.nyserda.org)

- This site features many energy efficiency programs and incentives geared toward commercial and industrial buildings.

Our Green Journey (www.galleyecocapital.com)

- A very savvy newsletter about green commercial real estate and finance, written by Lisa Galley, one of the real experts in this field. Well written, timely, and brief.

Tax Incentives Assistance Project (TIAP, www.energytaxincentives.org)

- Covers all federal tax incentives under the Energy Policy Act of 2005 and its subsequent updates.

U.S. Green Building Council (USGBC, www.usgbc.org)

- The premier Web site not only for the organization that created the LEED rating system, but also for news and happenings in the broader field of green buildings. If a trend has "legs," you'll find it here. You can download copies of all LEED rating systems and also search for LEED-registered and certified projects, as well as purchase the LEED-EBOM Reference Guide.

Note: Page numbers referencing figures are followed by an "*f*"; page numbers referencing tables are followed by a "*t*".

Verification of Sustainable Product Attributes

The growth of the number of new products and innovative approaches that are introduced into the building construction industry on a continuing basis has been phenomenal. In recent years, a large number of such new products or practices claim to be "green" or consistent with and promoting the goals of sustainable construction. Green building construction and green products are on a steep rise. The value of green building construction is projected to increase to $60 billion by 2010 (Source: McGraw-Hill Construction, 2008—Key Trends in the European and U.S. Construction Marketplace: Smart Market Report. U.S. Green Building Council: Green Building by the Numbers, November 2008) and the green building products market is projected to be worth $30-$40 billion annually by 2010 (Source: Green Building Alliance, 2006—Green Building Products: Positioning Southwestern Pennsylvania as the U.S. Manufacturing Center. U.S. Green Building Council). This issue becomes more complicated considering the global market and the fact that many products manufactured in one country find their way to multiple countries around the globe. Because it is rather impossible for designers, contractors or code officials to verify the credibility of each and every such claim regarding sustainable attributes, it is of paramount importance for everyone who is impacted by buildings, i.e. general population, as well as all involved in the construction industry, to be able to have a reliable source of verification of such claims. This requires a methodology and a standardized process by which to evaluate the degree of "greenness" and sustainable attributes of construction materials, elements and assemblies to result in buildings' green performance.

The most credible program available to address the issue of green verification is a program established by the International Code Council Evaluation Service (ICC-ES) known as the Sustainable Attributes Verification and Evaluation™ (SAVE™). The ICC-ES SAVE™ Program provides independent verification of manufacturers' claims about the sustainable attributes of their products. Successful evaluation under this program results in a Verification of Attributes Report™ (VAR™). The VAR™ can be helpful to qualify for points under major green rating systems or green codes and standards. The sustainable attributes are evaluated based on nine established guidelines. The current nine guidelines are:

- Evaluation Guideline for Determination of Recycled Content of Materials (EG101)
- Evaluation Guideline for Determination of Biobased Material Content (EG102)
- Evaluation Guideline for Determination of Solar Reflectance, Thermal Emittance and Solar Reflective Index of Roof Covering Materials (EG103)
- Evaluation Guideline for Determination of Regionally Extracted, Harvested or Manufactured Materials or Products (EG104)
- Evaluation Guideline for Determination of Volatile Organic Compound (VOC) Content and Emissions of Adhesives and Sealants (EG105)
- Evaluation Guideline for Determination of Volatile Organic Compound (VOC) Content and Emissions of Paints and Coatings (EG106)
- Evaluation Guideline for Determination of Volatile Organic Compound (VOC) Content and Emissions of Floor Covering Products (EG107)
- Evaluation Guideline for Determination of Formaldehyde Emissions of Composite Wood and Engineered Wood Products (EG108)
- Evaluation Guideline for Determination of Certified Wood and Certified Wood Content in Products (EG109)

The reports issued through the ICC-ES SAVE Program are similar to the concept of ICC-ES Evaluation Reports that have been used by the construction industry for several decades. These reports cover products that are alternatives to code-specified, provide independent evaluation of manufacturer claims, are broadly accepted by building officials, streamline the job of green verifiers, clarify environmental attributes of products and provide optional references to overall code compliance (i.e., structural, fire, durability, etc) via product ICC-ES Evaluation Reports.

The ICC-ES SAVE reports (VAR) are available online for access by anyone. A sample VAR is shown.

Most Widely Accepted and Trusted

ICC-ES SAVE Verification of Attributes Report™

VAR-1053

Issued July 1, 2009

This report is subject to re-examination in one year.

www.icc-es.org/save | 1-800-423-6587 | (562) 699-0543 *A Subsidiary of the International Code Council®*

DIVISION 07—THERMAL AND MOISTURE PROTECTION
Section 07 21 16—Building Insulation
Section 07210—Building Insulation

REPORT HOLDER:

A–1 Insulation, Inc.
123A Rocky Road
Asphalt, CA 43210
(123) 765-4321
www.a1insulation.com
jv@a1insulation.com

EVALUATION SUBJECT:

A–1 Insulation

1.0 EVALUATION SCOPE

Compliance with the following evaluation guideline:

ICC-ES Evaluation Guideline for Determination of Biobased Material Content (EG102), dated October 2008.

2.0 USES

A–1 Insulation is a semirigid, low-density, cellular isocyanate foam plastic insulation that is spray-applied as a nonstructural insulating component of floor/ceiling and wall assemblies.

3.0 DESCRIPTION

A–1 Insulation is a two component system with a nominal density of 1.0 pcf (16 kg/m^3). The insulation is produced by combining the two components on-site. Water is used as the blowing agent and reacts with the isocyanate, which releases a gas, causing the mixture to expand. The mixture is spray-applied to the surfaces intended to be insulated.

The insulation contains the minimum percentage of biobased content as noted in Table 1.

4.0 CONDITIONS

Evaluation of A–1 Insulation for compliance with the International Codes is outside the scope of this evaluation report. Evidence of compliance must be submitted by the permit applicant to the Authority Having Jurisdiction for approval.

5.0 IDENTIFICATION

The A–1 Insulation spray foam insulation described in this report is identified by a stamp bearing the manufacturer's name and address, the product name, and the VAR number (VAR-1053).

TABLE 1 – BIOBASED MATERIAL CONTENT SUMMARY

% MEAN BIOBASED CONTENT	METHOD OF DETERMINATION
15% (+/–3%)[1]	ASTM D6866

[1]Based on precision and bias cited in ASTM D 6866.

SAMPLE

ICC-ES Verification of Attributes Reports are issued under the ICC-ES Sustainable Attributes Verification and Evaluation Program (SAVE). These reports are not to be construed as representing aesthetics or any other attributes not specifically addressed, nor are they to be construed as an endorsement of the subject of the report or a recommendation for its use. There is no warranty by ICC Evaluation Service, Inc., express or implied, as to any finding or other matter in this report, or as to any product covered by the report.

Copyright © 2009